MONTESCANO (PAVIA) ITALY

Bi & Gi Publishers

Current Topics in Rehabilitation
Series Editor : R. Corsico M.D.

Titles in the series:

Respiratory Muscles in Chronic Obstructive Pulmonary Disease
Edited by: A. Grassino, C. Fracchia, C. Rampulla, L. Zocchi

Pathophysiology and Treatment of Pulmonary Circulation
Edited by: A. Morpurgo, R. Tramarin, C. Rampulla, C. Fracchia, F. Cobelli

Chronic Pulmonary Hyperinflation
Edited by: A. Grassino, C. Rampulla, N. Ambrosino, C. Fracchia

Forthcoming titles in the series:

Nutrition and Ventilatory Function
Edited by: R.D. Ferranti, C. Rampulla, C. Fracchia, N. Ambrosino

Right Ventricular Hypertrophy and Function in Chronic Lung Disease
Edited by: V. Jezeck, M. Morpurgo, R. Tramarin

Biochemistry of Pulmonary Emphysema
Edited by: C. Grassi, J. Travis, L. Casali

Acknowledgments: The Organizing Committee of the Workshop on *"Chronic Pulmonary Hyperinflation"* held in Montescano (Pavia) on 18-19, November 1988, gratefully acknowledges the Camillo Corvi S.p.A. Piacenza (Italy) for their support and co-operation.

Chronic Pulmonary Hyperinflation

Edited by: A. Grassino, C. Rampulla,
              N. Ambrosino, C. Fracchia

Foreword by R. Corsico
With 53 figures and 12 tables

Springer-Verlag Italia Srl.

A. Grassino   - Meakins-Christie Laboratories, McGill University and Notre Dame Hospital,
                    University of Montreal, Quebec, Canada
C. Rampulla   - Medical Center of Rehabilitation, Montescano, Pavia, Italy
N. Ambrosino   - Medical Center of Rehabilitation Montescano, Pavia, Italy
C. Fracchia   - Medical Center of Rehabilitation, Montescano, Pavia, Italy

Series Editor:
R. Corsico -   Medical Center of Rehabilitation, Montescano, Pavia, Italy

ISBN 978-1-4471-3784-9   ISBN 978-1-4471-3782-5 (eBook)
DOI 10.1007/978-1-4471-3782-5

British Library Cataloguing in Publication Data
Chronic pulmonary hyperinflation
1. Humans: Respiratory systems. Diseases
I. Grassino, A. Alex II. Current topics in rehabilitation 616.2

Library of Congress Cataloging-in-Publication Data
Workshop on "Chronic Pulmonary Hyperinflation" (1988: Montescano, Italy)
Chronic Pulmonary Hyperinflation / edited by A. Grassino... (et al.): foreword by R. Corsico.
(Current topics in rehabilitation)
"Workshop on: "Chronic Pulmonary Hyperinflation" held in Montescano (Pavia) on November
18-19, 1988" - Ser, t.p. Includes index

1. Lungs--Diseases, Obstructive--Congresses. 2. Lungs--Pathophysiology--Congresses. I. Grassino,
A. II. Title, III Series.
(DNLM: 1. Lung Diseases, Obstructive--physiopathology--congresses. 2. Lung Diseases, Obstruc-
tive--rehabilitation. WF 600 W9253c 1988) RC776.03W65 1988
616.2' 4--dc20
DNLM/DLC for Library of Congress 90-10451 CIP

© 1991 Springer-Verlag Italia
Softcover reprint of the hardcover 1st edition 1991
Originally published by Bi & Gi Publisher s.r.l in 1991

Typeset by Bi & Gi, Verona, Italy

Foreword

The workshops that have been held over the past few years and the volumes published in their wake have proved highly successful and have prompted us to press on with our initial plans.

Our basic aim was to tackle certain very important problems in respiratory rehabilitation and then discuss the various issues with people from all over the world engaged in the updating of experience and knowledge in this field.

We therefore firmly believe that this ongoing effort is of fundamental importance.

Hyperinflation, which is still a poorly defined clinical and physiopathological condition, is the focal point of this present study, which is aimed at discussing and weighing up the physiopathological mechanisms, clinical consequences, and rehabilitation possibilities in a disease in which, until relatively recently, rehabilitation had seemed almost totally ineffective.

The present contributions, however, show us how very important and versatile rehabilitation may be in its treatment.

Perhaps, if we consider this branch of medicine as one which now no longer simply draws upon other sciences, but make an active contribution in its own right, we will have touched upon the most important aspect of this study.

If I may, I would just like to add how very pleasing it is for me to acknowledge how much this joint effort has contributed, in terms of true insights and above all results, to finding solutions to the problems addressed in recent years.

These results have in turn spurred us on to continue the invaluable work, the next major theme on our agenda being "Nutrition and Ventilatory Function".

I would therefore like once again to take this opportunity of thanking all the contributors, without whose skill and commitment this book would not have been possible.

November 1989

RENATO CORSICO

Preface

The focus of this book is on chronic pulmonary hyperinflation, its causes, consequences and treatment.

This condition is a direct cause of airways and interstitial disease of the lung and at the same time is the cause of changes in chest wall mechanics leading to respiratory muscle dysfunction and eventually to their inability to sustain normal ventilation.

Lung hyperinflation is also a prime example of systems interdependence, where one pathology triggers related system abnormalities.

The initial chapters deal with the basic mechanisms leading to alveolar wall disruption, with increase in airways resistance and air trapping. These are followed by a series of chapters which explore the effects of hyperinflation and gas mixing on lung and bronchial circulation, and on venous returns to the heart, and its role in pulmonary edema.

Subsequent chapters explore how lung hyperinflation, both acute and chronic, can affect inspiratory muscle function and how tonic activity of these muscles can in turn be responsible for hyperinflation.

The final chapters deal with therapeutic aspects and include such topics as mechanical ventilation, bronchodilatation, physiotherapy and respiratory muscle training in patients with chronic hyperinflation.

The authors of the articles in this book were invited to attend an international workshop on the basis of the excellence of their past contributions to the literature and their original views and experience in this field.

Many other participants from Europe and North America contributed considerably to lively discussions, unfortunately not included here.

This workshop was largely the product of the dedicated and skilful organizational work of Drs. C. Rampulla, N. Ambrosino and C. Fracchia, and the continuous support of the Fondazione Clinica del Lavoro S. Maugeri to projects aimed at promoting education in the field of medical rehabilitation.

ALEX E. GRASSINO

Contents

Contributors

AGOSTONI P.
Institute of Cardiology, Institute of Cardiovascular Research "G. Sisini", C.N.R.,
Cardiological Center Fondazione I. Monzino", University of Milan, Italy
AMBROSINO N.
"S. Maugeri Foundation" Pavia, Care and Research Institute, Medical Center of Rehabi-
litation, IRCCS, Fondazione Clinica del Lavoro, Montescano, Pavia, Italy
APPELBAUM A.
Department of Cardiothoracic Surgery, Hadassah University Hospital, Jerusalem, Israel
BADIER M.
Respiratory Physiology Laboratory, CNRS, Faculty of Medicine, Marseille, France
BANAUDI C.
Respiratory Physiopathology Service, S. Luigi Hospital, Orbassano, Turin, Italy
BEGIN P.
Meakins-Christie Laboratories, McGill University and Notre Dame Hospital, University
of Montreal, Quebec, Canada
BELLIA V.
Respiratory Physiopathology Institute, National Research Council, Pneumology Depart-
ment, University of Palermo, Italy
BERNASCONI M.
Department of Anesthesia and Intensive Care, City Hospital, Institute of Occupational
Medicine, University of Padua, Italy
BOARO D.
Respiratory Physiopathology Service, S. Luigi Hospital, Orbassano, Turin, Italy
BONANDRINI L.
Chair of Microsurgery, University of Pavia, Italy
BONMARCHAND G.
Respiratory Physiopathology Group, Ch. Nicolle Hospital, Rouen, France
BONSIGNORE G.
Respiratory Physiopathology Institute, National Research Council, Pneumology Depart-
ment, University of Palermo, Italy
BORGHETTI A.
Institute of Clinical Medicine and Nephrology, University of Parma, Italy
BRANDOLESE R.
Department of Anesthesia and Intensive Care, City Hospital, Institute of Occupational
Medicine, University of Padua, Italy
CAMINITI G.
Pneumology Department, Saint-Pierre University Hospital, Brussels, Belgium

CARMIGNANI G.
CNR Institute of Clinical Physiology and 2nd Division of Internal Medicine, University of Pisa, Italy
CARROZZI L.
CNR Institute of Clinical Physiology and 2nd Division of Internal Medicine, University of Pisa, Italy
CIACCIA A.
Institute of Pulmonary Diseases, University of Ferrara, Italy
CIBELLA F.
Respiratory Physiopathology Institute, National Research Council, Pneumology Department, University of Palermo, Italy
COFFRINI E.
Institute of Clinical Medicine and Nephrology, University of Parma, Italy
DAL VECCHIO L.
Department of Anesthesia and Intensive Care, City Hospital, Institute of Occupational Medicine, University of Padua, Italy
DECRAMER M.
Respiratory Department, University Hospital, Catholic University, Leuven, Belgium
DERENNE J. PH.
Pneumology Service, Saint-Antoine Hospital, Paris, France
DI PEDE F.
CNR Institute of Clinical Physiology and 2nd Division of Internal Medicine, University of Pisa, Italy
DONNER C.F.
"S. Maugeri Foundation" Pavia, Medical Center of Rehabilitation, Fondazione Clinica del Lavoro, Veruno, Novara, Italy
DORIA E.
Institute of Cardiology, Institute of Cardiovascular Research "G.Sisini", C.N.R., Cardiological Center "Fondazione I. Monzino", University of Milan, Italy
DREYFUSS D.
Medical Resuscitation Service, Louis Mourier Hospital, Colombes, INSERM U82, Faculty of Xavier Bichat, Paris, France
EMERY C.
Department of Medicine, Royal Hallamshire Hospital, University of Sheffield, England
FABBRI L. M.
Laboratory of Respiratory Pathophysiology, Institute of Occupational Medicine, University of Padua, Italy
FIACCADORI E.
Institute of Clinical Medicine and Nephrology, University of Parma, Italy
FITTING J.W
Pneumology Department, Internal Medicine Department, Vaudois University Hospital, Lausanne, Switzerland
FRACCHIA C.
"S. Maugeri Foundation" Pavia, Care and Research Institute, Medical Center of Rehabilitation, IRCCS, Fondazione Clinica del Lavoro, Montescano, Pavia, Italy
GIUNTINI C.
CNR Institute of Clinical Physiology and 2nd Division of Internal Medicine, University of Pisa, Italy

GRASSINO A.
Meakins-Christie Laboratories, McGill University and Notre Dame Hospital, University of Montreal, Quebec, Canada

GROSS D.
Department of Anesthesiology, Hadassah University Hospital, Jerusalem, Israel

GUARIGLIA A.
Institute of Clinical Medicine and Nephrology, University of Parma, Italy

GULOTTA C.
Respiratory Physiopathology Service, S. Luigi Hospital, Orbassano, Turin, Italy

HOWARD P.
Respiratory Function Unit, Department of Medicine, Royal Hallamshire Hospital, University of Sheffield, England

IOLI F.
"S. Maugeri Foundation" Pavia, Medical Center of Rehabilitation, Fondazione Clinica del Lavoro, Veruno, Novara, Italy

JAMMES Y.
Respiratory Physiology Laboratory, CNRS, Faculty of Medicine, Marseille, France

LACHMAN A.
Pneumology Department, Saint-Pierre University Hospital, Brussels, Belgium

LANDUCCI L.
CNR Institute of Clinical Physiology and 2nd Division of Internal Medicine, University of Pisa, Italy

MACALUSO C.
Respiratory Physiopathology Institute, National Research Council, Pneumology Department, University of Palermo, Italy

MAMMINI U.
CNUCE, University of Pisa, Italy

MILIC-EMILI J.
Department of Physiology, Meakins-Christie Laboratories, McGill University, Montreal, Quebec, Canada

MONTAGNA T.
"S. Maugeri Foundation" Pavia, Care and Research Institute, Medical Center of Rehabilitation, IRCCS, Fondazione Clinica del Lavoro, Montescano, Pavia, Italy

NAVA S.
"S. Maugeri Foundation" Pavia, Care and Research Institute, Medical Center of Rehabilitation, IRCCS, Fondazione Clinica del Lavoro, Montescano, Pavia, Italy

ORCEL B.
Pneumology Service, Saint-Antoine Hospital, Paris, France

PAOLETTI P.
CNR Institute of Clinical Physiology and 2nd Division of Internal Medicine, University of Pisa, Italy

PAPI A.
Institute of Pulmonary Diseases, University of Ferrara, Italy

PATESSIO A.
"S. Maugeri Foundation" Pavia, Medical Center of Rehabilitation, Fondazione Clinica del Lavoro, Veruno, Novara, Italy

PEPI M.
Institute of Cardiology, Institute of Cardiovascular Research "G.Sisini", C.N.R., Cardiological Center "Fondazione I. Monzino", University of Milan, Italy

PIPITONE P.
Respiratory Physiopathology Institute, National Research Council, Pneumology Department, University of Palermo, Italy

POGGI R.
Department of Anesthesia and Intensive Care, City Hospital, Institute of Occupational Medicine, University of Padua, Italy

POZZOLI M.
"S. Maugeri Foundation" Pavia, Care and Research Institute, Medical Center of Rehabilitation, IRCCS, Fondazione Clinica del Lavoro, Montescano, Pavia, Italy

PRITCHARD S.M.
Respiratory Function Unit, Department of Medicine, Royal Hallamshire Hospital, University of Sheffield, England

RAMPULLA C.
"S. Maugeri Foundation" Pavia, Care and Research Institute, Medical Center of Rehabilitation, IRCCS, Fondazione Clinica del Lavoro, Montescano, Pavia, Italy

RONDA N.
Institute of Clinical Medicine and Nephrology, University of Parma, Italy

ROSSI A.
Department of Anesthesia and Intensive Care, City Hospital, Institute of Occupational Medicine, University of Padua, Italy

SAETTA M.
Laboratory of Respiratory Pathophysiology, Institute of Occupational Medicine, University of Padua, Italy

SAUMON G.
Medical Resuscitation Service, Louis Mourier Hospital, Colombes, INSERM U82, Faculty of Xavier Bichat, Paris, France

SERA G.
Respiratory Physiopathology Service, S. Luigi Hospital, Orbassano, Turin, Italy

SERGYSELS R.
Pneumology Department, Saint-Pierre University Hospital, Brussels, Belgium

TAMBORINI G.
Institute of Cardiology, Institute of Cardiovascular Research "G.Sisini", C.N.R., Cardiological Center "Fondazione I. Monzino", University of Milan, Italy

TARDIF C.
Respiratory Physiopathology Group, Ch. Nicolle Hospital, Rouen, France

TORBICKI A.
Department of Hypertension and Angiology, Medical Academy, Warsaw, Poland

TORCHIO R.
Respiratory Physiopathology Service, S. Luigi Hospital, Orbassano, Turin, Italy

TOSADORI A.
Respiratory Physiopathology Service, S. Luigi Hospital, Orbassano, Turin, Italy

TRAMARIN R.
"S. Maugeri Foundation" Pavia, Care and Research Institute, Medical Center of Rehabilitation, IRCCS, Fondazione Clinica del Lavoro, Montescano, Pavia, Italy

VIEGI G.
CNR Institute of Clinical Physiology and 2nd Division of Internal Medicine, University of Pisa, Italy

VITALI P.
Institute of Clinical Medicine and Nephrology, University of Parma, Italy
WATERHOUSE J.C.
Respiratory Function Unit, Department of Medicine, Royal Hallamshire Hospital, University of Sheffield, England
WILLEPUT R.
Pneumology Department, Saint Pierre University Hospital, Brussels, Belgium
ZOCCHI L.
"S. Maugeri Foundation" Pavia, Care and Research Institute, Medical Center of Rehabilitation, IRCCS, Fondazione Clinica del Lavoro, Montescano, Pavia, Italy

Basic Mechanisms

1. Distribution of Hyperinflation in a General Population

P. Paoletti[1], G. Viegi[1], G. Carmignani[1], L. Carrozzi[1], U. Mammini[2],
L Landucci[1], F. Di Pede[1], C.Giuntini[1]
1. CNR Institute of Clinical Physiology and 2nd Division of Internal Medicine, University of Pisa, Italy
2. CNUCE, University of Pisa, Italy

Introduction

Hyperinflation is usually related to the clinical stages of chronic obstructive lung disease (COLD) and it specifically characterizes these patients.[1]

The presence of hyperinflation in the pre-clinical stage of COLD has not been investigated, mainly because of the difficult and time-consuming methods (multiple breath dilution methods and body plethysmography) used to measure the functional parameters characterizing this condition in population based studies.

We had the opportunity to use the single breath helium (He) dilution method to measure carbon monoxide (CO) diffusing capacity and total lung capacity in a general population sample enrolled in a prospective study to investigate the natural history of COLD.

The aim of this report is to describe the distribution of some parameters characterizing hyperinflation in a sample of a general population and their association with cigarette smoking.

Material and Methods

3289 subjects living in the rural, unpolluted area of the Po River Delta (North Italy, 30 kilometers south of Venice) were enrolled in a longitudinal study on the natural history of COLD and to evaluate the long term effects of sulphur dioxide exposure emanating from a large thermoelectric power plant.

The characteristics of the population sample, the experimental design the prevalence of symptoms and disease, and the reference equations for lung function parameters have been previously reported.[2-6]

Briefly the population sample (8 - 64 years) was selected using a multistage stratified family cluster design. Agricultural, trading and fishing activities are mostly represented in the area; pollution from industrial production as well as from large cities is absent in the area.

A modified NHGLI administered questionnaire[7] was used to assess the presence of respiratory symptoms, diseases and the risk factors for COLD.

Lung function measurements were obtained, using standard protocols developed by the Special Project on Chronic Lung Disease of the Italian Research Council. Protocols were based essentially on the recommendations of the American Thoracic Society (ATS) with some exceptions.[8] Subjects performed the lung function tests in the following sequence:

    1. slow vital capacity (VC);

    2. single breath nitrogen test (SBN2);

    3. single breath diffusing capacity for carbon monoxide (DLCOsb);

    4. forced vital capacity (FVC) and derived indices (FEV_1, FEF and Vmax).

Details on the equipment, performance and criteria of acceptability have been previously reported.[3-4]

Total lung capacity from the single breath helium dilution (TLCsb) was computed during the standard maneuver to obtain the DLCOsb. An automated instrument (Hewlett-Packard 47804/S) was used: volume was measured by a pneumotachograph (Fleisch n.2) after flow integration and helium was measured by a thermoconductivity He analyzer. Signals were fed to a computer for automated analyses. A demand valve was in connection with a tank containing 3% CO_2, 10% He, 20% O_2 and the balance was nitrogen. Gas analyzers were calibrated before testing each subject and the pneumotachograph was calibrated daily and checked by a 3 liter syringe. The sequence of the in-expiratory maneuvers to obtain TLCsd was the standard maneuver suggested by the ATS to obtain DLCOsb: after a few minutes of adaptation, subjects exhaled rapidly to residual volume (RV) and then they rapidly inspired the gas mixture from the tank to total lung capacity (TLC). Subjects held their breath for 10 seconds and then exhaled rapidly to RV, while the valve system allowed the collection of 1 liter of gas in a bag for the analyses, after the elimination of the dead space. TLCsb was computed using the standard formula from the single breath He dilution and was expressed at BTPS:[3]

TLCsb = IVC ATPD x FIHe/FAHe x (AIPD-BTPS).

RV was computed after subtracting VC previously measured from TLCsb; if the value of FVC was higher than VC, the former was used to subtract from TLCsb. Finally, the RV/TLC% ratio was computed to obtain another index to characterize hyperinflation. 2688 subjects (80%) were able to perform the test; subjects older than 20 years (904 males and 956 females) were considered for the analyses.

Statistical analyses were performed at the University of Pisa Computer Center (CNUCE), using the Statistical Package for Social Sciences (SPSSx) routines. Analysis of variance (ANOVA), covariance and regression analysis were used.

Results

Reference values for RV were derived after the selection of "normal" subjects using strict criteria.[3-4] Prediction equations were computed considering two age groups sepparately to take into account the cross-sectional effect due to growth in young ages (paper submitted for publication).

In Fig. 1 RV percent predicted (RV%) by age for smokers and non smokers > 20 years of both sexes is shown: RV% increases with age and the increase is higher in smokers of both sexes (the correlation is significant only in smokers, $r = 44$, $p > .001$, females, $r = 29$; $p < .03$).

RV % BY AGE IN MALES AND IN FEMALES > 20 YEARS

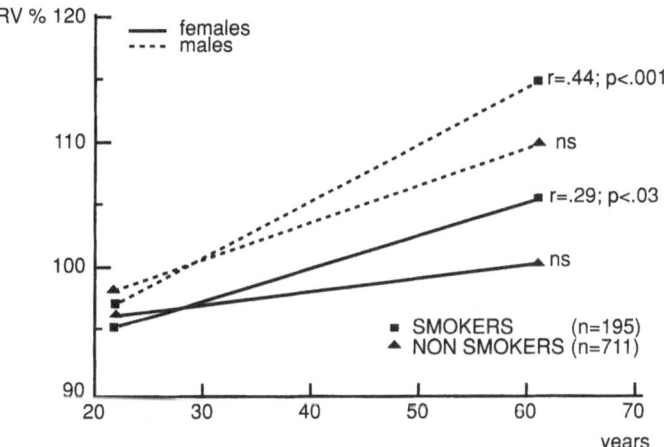

Fig. 1. RV% by age in smokers and non smokers: males and females. The slope is steeper in smokers of both sexes and the correlation is significant.

In Fig. 2, mean values of RV% and RV/TLC% are reported for smokers and non smokers > 20 years of both sexes. Mean values are adjusted for age because of the significant relation of these parameters with age in males, both RV% and RV/TLC% are higher in smokers (10% and 32% vs 101% and 30% in non smokers, respectively); however, the difference was not significant. In females only RV/TLC% was higher in smokers (not significant).

Then, we have evaluated the effect of different smoking exposure, grouping subjects according to the number of pack years, an index which takes into account the duration and the quantity of the exposure (n°of cigarette per day x n° of years

6

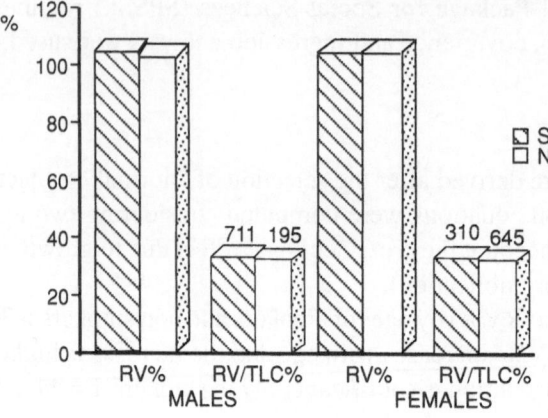

RV% AND RV/TLC% BY SMOKING-MALES AND FEMALES > 20 YEARS

SMOKERS
NON SMOKERS

711 195   310 645

MALES   FEMALES

Fig. 2. Mean values of RV% and RV/TLC% in smokers (hatched column) and non smokers (white column) of both sexes. The difference is not significant by ANOVA.

of smoking/20). In Fig. 3, age adjusted RV% and RV/TLC% are reported for smokers of both sexes according to the number of pack years they have smoked: males < 9 pack years (n = 182), 9-21 pack years (n = 266), > 21 pack years (n = 261); females < 4 pack years (n = 96), 4 - 6 pack years (n = 94), > 6 pack years (n = 119). In males with > 21 pack years, RV% and RV/TLC% showed higher values than those with lower pack years: 110%, 105% and 102%, 36%, 29% and 25%, respectively (significant at p < .01 by ANOVA only for RV/TLC%). In females, differences among the groups were not present (a slight trend is present for RV/TLC%).

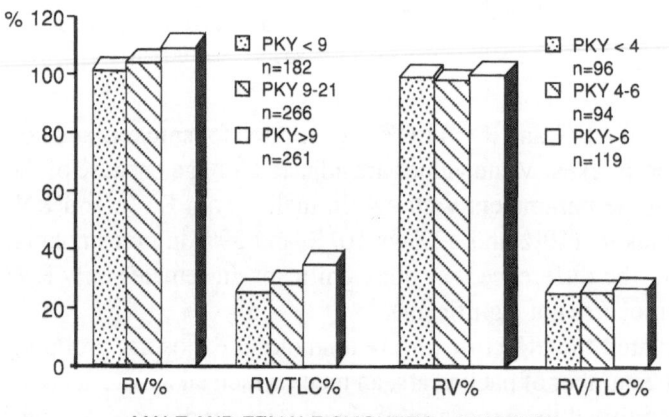

RV% AND RV/TLC% BY PACK-YEARS

PKY < 9
n=182
PKY 9-21
n=266
PKY>9
n=261

PKY < 4
n=96
PKY 4-6
n=94
PKY>6
n=119

MALE AND FEMALE SMOKERS > 20 YEARS

Fig. 3. Age adjusted mean values of RV% and RV/TLC% in smokers, after having grouped the subjects according to pack years. Left side: in males, a dose-response effect is present and subjects with > 21 pack years have significantly higher values of RV/TLC% (by ANOVA). Right side: in females, RV% and RV/TLC% are similar among the three groups.

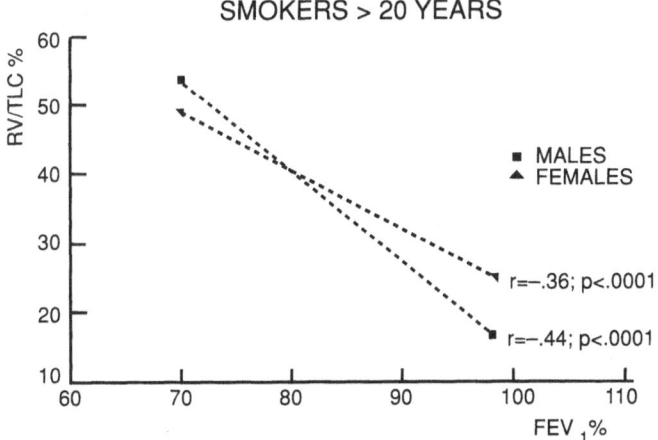

Fig. 4. Linear regression of RV/TLC% by FEV$_1$% in smokers of both sexes: a negative significant correlation is present.

Finally, in Fig. 4 the relationships between FEV$_1$% predicted and RV/TLC% are reported in smokers of both sexes: significant negative correlations are present (males: r = - 48; p < .0001; females: r = - 36; p < .0001). The same result was present in non smokers (males: r = - 42; p < .0001; females: r = - 35; p < .0001).

Discussion

The cross-sectional distribution of residual volume (expressed in percent predicted) shows an increase with age. This volume increase may be related to the anatomical changes of the ageing lung, mainly caused by the loss of the elastic recoil.[9-10] The ageing effect is more evident in smokers both in males and females, underlining the important contribution of the adverse effect of cigarette smoking on the respiratory tract.

By grouping the subjects as smokers and non smokers, age adjusted mean value of RV% and RV/TLC% were not significantly different between the two groups of both sexes, however, RV/TLC% was higher in smokers when the same analysis was performed only in older male smokers (>40 years), the difference being significant. This finding indicates that older smokers, beside having a higher RV because of the ageing effect, were exposed to a higher cigarette consumption. In addition, the analysis shown in Fig.3 indicates a dose-response increment of RV% and RV/TLC%, especially in males. In females the trend is evident only for RV/TLC% and the dose-response effect is not present, probably because of the low cigarette consumption of female smokers in our population.[5] The increase in RV% in smokers may be ascribed to an increased airway obstruction and probably to initial emphysema. Indeed, a significant association between airway obstruction and increased residual volume was observed in the analysys shown in Fig. 4, suggesting that the mechanism of air trapping may be responsible for the increased residual volume. Peripheral airway inflammation is a common finding in smokers (with or without mucus hypersecretion)[11-12] and these pathological changes may explain the increase in RV. It is more difficult to relate the increase in RV to the presence of initial changes related to emphysema.

In fact, in order to achieve a precise diagnosis of emphysema, chest radiographs should be performed with the addition of chest computerized tomography, as recently suggested by Gould et al.[13] Both these assessments are unsuitable in epidemiological studies.

Methods based on single breath helium dilution to measure total lung capacity have been criticized because of the lack of precision, especially in subjects with COLD.[14] However, while TLCsb in patients with severe obstruction is under evaluated because of the presence of a higher degree of dyshomogeneous distribution of ventilation, in normal subjects TLCsb is similar to that measured by the rebreathing helium technique.[15] In our population few subjects showed a severe obstruction, and reference values for TLCsb were derived in "normal" subjects using strict criteria of selection.[3-4] Indeed, our reference equations for DLCOsb were among the highest published in the literature[16-17] and, among the different explanations, one of the reasons is the high value of TLCsb. Consequently, TLCsb used to derive RV is quite accurate and certainly not under estimated.

It was interesting to find that higher values of RV% and RV/TLC% were observed in those subjects with lower values of FEV_1%. This observation suggests that in subjects (mostly smokers) with initial obstruction TLCsb is not underestimated and that the reduction of VC may explain the relative increase in RV.

In conclusion, our data confirm the importance of using the CO single breath diffusing capacity in epidemiological studies to obtain not only the diffusing capacity but also reliable measurements of TLCsb, which may be used to derive residual volume and to detect initial hyperinflation.

References

1. Burrows B., Fletcher C.M., Heard B.E, Jones N.L., Wootliff J.S.: The emphysematous and bronchial types of chronic airway obstruction. A clinicopathological study of patients in London and Chicago. Lancet, 1966; 1: 830-835.
2. Carrozzi L., Giuliano G., Viegi G., Paoletti P., Di Pede F., Mammini U., Saracci R., Giuntini C., Lebowitz M.D.: The Po River Delta epidemiological study of obstructive lung disease: sampling methods, environmental and population characteristics. Eur. J. Epidemiol. (in press).
3. Paoletti P., Viegi G., Pistelli G., Di Pede F., Fazzi P., Polato R., Saetta M., Zambon R., Carli G., Giuntini C., Lebowitz M.D., Knudson R.J: Reference equation for the single breath diffusing capacity: a cross sectional analysis and effect of body size and age. Am. Rev. Respir. Dis. 1985; 132: 806-813.

4. Paoletti P., Pistelli G., Fazzi P., Viegi G., Di Pede F., Prediletto R., Carrozzi L., Polato R., Saetta M., Zambon R., Sapigni T., Lebowitz M.D., Giuntini C.: Reference values for vital capacity and flow volume curves from a general population. Bull. Eur. Physiopathol. Respir. 1986; 22: 451-456.

5. Viegi G. Paoletti P., Prediletto R., Carrozzi L., Fazzi P., Di Pede F., Pistelli G., Giuntini C., Lebowitz M.D.: Prevalence of respiratory symptoms in an unpolluted area of North Italy. Eur. Respir. J. 1988; 1: 311-318.

6. Viegi G., Paoletti P., Di Pede F., Prediletto R., Carrozzi L., Pistelli G., Giuntini C.: Single breath nitrogen test in an epidemiological survey reliability, reference values and relationships with symptoms. Chest 1988; 93: 1213-1220.

7. Fazzi P., Viegi G., Paoletti P., Giuliano G., Begliomini E., Fornal E., Giuntini C.: Comparison between two standardized questionnaires and pulmonary function tests in a group of workers. Eur. J. Respir. Dis. 1982; 63: 168-169.

8. Pistelli G., Carmignani G., Paoletti P., Di Pede F., Viegi G., Carrozzi L., Celi A., Giuntini C.: A comparison of algorithms for determination of end-point of the forced vital capacity. Chest 1987; 91: 100-105.

9. Knudson R.J., Clark D.F., Kennedy T.C. et al: effect of aging along on the mechanical properties of the normal adult human lung. J. Appl. Physiol. 1977; 43: 1054-1062.

10. Colebatch H.J.H, Greaves IA, Ng CKY: Exponential analysis of elastic recoil and aging in healthy males and females. J. Appl. Physiol. 1979; 47: 638-691.

11. Niewoehner D.E., Kleinerrman J., Rice F., Pathological changes in the peripheral airways of young cigarette smokers. N. Engl. J. Med. 1974; 292: 755-758.

12. Thurlbeck W.M., Chronic airflow obstruction in lung disease. W.B. Saunders Philadelphia, 1976.

13. Gould G.A., Macnee W., Warren P.M., Redpath A., Best J.J.K, Lamb D., Flenley DC.: CT measurements of lung density in life can quantitate distal airspace enlargement-an essential defining feature of human emphysema. Am. Rev. Respir. Dis. 1988; 137: 380-392.

14. Teculescu D.B., Stanescu D.C.: Total lung capacity in obstructive lung disease. Comparative determinations by single-and multiple-breath helium dilution. Bull. Eur. Physiopathol. Respir. 1969; 5: 453-464.

15. George R., Saumon G., Loiseau A.: The relationship of age to pulmonary membrane conductance and capillary blood volume. Am. Rev. Respir. Dis. 1978; 117: 1069-1076.

16. Knudson R.J., Kalternborn W.T., Knudson D.E., Burrows B.: The single breath carbon monoxide diffusing capacity: reference equations derived from a healthy non-smoking population and effects of hematocrit. Am. Rev. Respir. Dis. 1987; 137: 805-811.

17. Paoletti P., Viegi G., Pistelli G., Di Pede F., Giuntini C. :Correspondence reference equations for the single breath diffusing capacity. Am. Rev. Respir. Dis. 1986; 133: 1210-1211.

2. Pathology and Biochemical Basis of Chronic Pulmonary Hyperinflation

M. Saetta[1], L.M. Fabbri[1], A. Papi[2], A. Ciaccia[2]
1. Laboratory of Respiratory Pathophysiology, Institute of Occupational Medicine, University of Padua, Italy
2. Institute of Pulmonary Diseases, University of Ferrara, Italy

Introduction

Chronic pulmonary hyperinflation, which is defined as a permanent increase in the end-expiratory lung volume above the predicted functional residual capacity, is a characteristic feature of emphysema. Chronic pulmonary hyperinflation has a static component and a dynamic component.

In normal subjects during tidal breathing the end-expiratory lung volume corresponds to the elastic equilibrium volume of the total respiratory system. Emphysema is associated with a decrease in lung elastic recoil which determines a marked increase in the functional residual capacity and in the elastic equilibrium volume.

This is the static component of chronic pulmonary hyperinflation. In emphysema, the end-expiratory lung volume is significantly above the elastic equilibrium volume, because a complete expiration is impaired by high respiratory resistances and expiratory flow limitation. This is the dynamic component of chronic pulmonary hyperinflation.

While many physiological aspects of chronic pulmonary hyperinflation have been deeply investigated in recent years,[1] the pathologic basis of the hyperinflated lung is still poorly understood.[2]

In the first part of this paper we will briefly illustrate some findings on the morphology of chronic pulmonary hyperinflation taken from our own experience;[3-4] in the second part we will provide a brief review of the pathogenesis of hyperinflation of the lung, and particularly the biochemical basis of irreversible hyperinflation, i.e. emphysema.

Morphology of Pulmonary Hyperinflation

The aim of our study was to quantify the morphology of the hyperinflated lung and to investigate a possible quantitative relationship between physiological and morphological changes in chronic pulmonary hyperinflation. As a model to quantify the morphology of chronic pulmonary hyperinflation we used lungs of smokers, since cigarette smoking is the overwhelming factor associated with emphysema. For more than two decades the definition of emphysema has been "a condition of the lung characterized by abnormal, permanent enlargement of airspaces distal to the terminal bronchiole, accompanied by the destruction of their walls". This definition was amended recently to exclude lungs in which enlargement and destruction of peripheral airspaces are associated with diffuse pulmonary fibrosis.[5] Therefore, there are, by definition, two components in emphysema, the enlargement of airspaces and the destruction of alveolar walls. As Thurlbeck pointed out,[2] the definition is clear but the precise application of it is not established. For example "abnormal" enlargement implies a quantitative phenomenon, and "destruction" has never been precisely defined and quantified.

The most widely used microscopic method to quantify airspace enlargement in emphysema is the mean linear intercept (Lm) that measures the average distance between alveolar walls.[2] The upper limit of Lm for 'normal' lungs is 0.350 mm and conventionally lungs with Lm within this value are considered to be free of emphysema. However, there are many problems regarding the usefulness of Lm in quantifying emphysema and the most important is that an increased dilatation of airspace without destruction leads to an increased Lm unrelated to the classical definition of emphysema. This is exemplified by the ageing lung, where Lm is increased without apparent destruction.[2] Therefore, the use of the dimensional component alone, which is the measurement of airspace enlargement, is not adequate to quantify emphysema. A better way would be to combine the airspace enlargement with a quantification of alveolar wall destruction. In fact, the destruction of alveolar walls is an important part of the definition of emphysema; however, it has never been well defined and quantified[2] and it is usually considered as a very final event of the disease. The only available methods to quantify destruction were macroscopic methods. But with these methods only obvious, advanced levels of destruction could be detected, whereas at a stage in which emphysema is not macroscopically evident, no methods were available to quantify destruction. We therefore devised two microscopic indices to quantify destruction in the overall parenchyma[3] and around the small airways.[4]

We examined lungs or lobes from smokers undergoing thoracotomy for localized pulmonary lesions, who performed pulmonary function tests, including static pressure-volume curve of the lung, one week before surgery. As a control group we studied autopsy lungs from non smokers who died suddenly and had no history of respiratory diseases. We found that smokers had an increased amount of destruction of alveolar walls both in the overall parenchyma and around the small airways compared to non smokers, and that this destruction was related to the inflammatory process in the walls of the bronchioles. It is likely that the inflammatory infiltrate of the bronchiolar wall may spread over the attached alveolar walls and that by products of the inflammatory cells may weaken alveolar tissue and may in turn facilitate its rupture, especially the point of attachment between the bronchiolar and alveolar wall, where the mechanical stress is probably maximum.

In smoker lungs we observed cases that, without airspace enlargement (Lm within 0.350 mm) already have a large amount of parenchymal destruction (increased destructive index). These findings support the hypothesis that destruction of lung tissue, instead of being a final event of the disease, can precede the airspace enlargement in smokers. We next investigated the functional significance of these early morphological abnormalities and we found a significant correlation between the parenchymal destruction (both around the airways and in the overall parenchyma) and the loss of elastic recoil of the lung (measured on the static pressure-volume curve that the smokers performed before surgery).

In conclusion loss of alveolar attachments and parenchymal destruction play an important role in both airway compressibility and loss of elastic recoil (dynamic and static hyperinflation) in smokers; these morphological changes are related to the inflammatory process of the lung and are present in the early stage of disease in smokers.

In the next section we will briefly review the biochemistry and function of the pulmonary interstitium, and the biochemical mechanisms that have been suggested as causing lung parenchymal destruction and thus chronic pulmonary hyperinflation.

Biochemistry of the Pulmonary Interstitium

The pulmonary interstitium consists of a continuous network that includes branches departing around the bronchial and pulmonary vascular tree and extending to the alveoli. The pulmonary interstitium is connected through the basement membranes to both epithelial and endothelial cells and it represents a continuum from the visceral pleura to the hilum. The pulmonary interstitium provides the support for the pulmonary architecture, is a major determinant of the mechanical and gas exchange properties of the lung, and modulates the passage of inflammatory cells, fluid and solutes through the lung. The pulmonary interstitium consists of the fibroelastic skeleton and of mesenchymal cells. The fibroelastic skeleton supports airways, blood vessels and alveoli. Basement membranes support endothelial and epithelial cells, and the pleural fibrillar network of collagen and elastic fibers function as a restraint for the lung expansion.[6-8]

Structure

The fibroelastic skeleton consists of the fibrillar components collagen (60-70%), and elastin (25-30%), and of the amorphous components glycosaminoglycans (1%) and fibronectin (0.5%).[9] In the normal lung, collagen is mainly types I and III, but also a small amount of type V is present. Elastin consists of tightly cross-linked tropoelastin molecules and glycoprotein microfibrils, which are the internal and external component of the elastic fibers respectively. Glycosaminoglycans are the polysaccharide component of proteoglycans, macromolecules containing a protein core linked to a large polysaccharide side chain.[9] In the normal lung glycosaminoglycans include dermatan sulfate, the main component of the alveolar interstitium, heparan sulfate, the main component of basement membranes, hyaluronic acid, chondroitin-4-sulfate, chondroitin-6-sulfate, and heparin. Fibronectin includes a group of glycoproteins.

Basement membrane and pleura may be considered specialized structures of the pulmonary interstitium. The biochemical structure of the basement membranes differs from the pulmonary interstitium. Its major components are collagen IV, mainly distributed in the lamina densa, laminin, mainly distributed in the lamina lucida, heparan sulfate, proteoglycan and nidogen.

These components are mainly synthesized by endothelial and epithelial cells. The visceral pleura consists of an outer layer of mesothelial cells resting on a collagen basement underneath which there is a network of elastic fibers connecting the pleura to the alveoli.

Mesenchymal cells, macrophages and lymphocytes are present in the alveolar interstitium. The mesenchymal cells represent 30-40% of the parenchymal cells present in the alveolar wall. They include fibroblasts and myofibroblasts, which are present in larger numbers, but also smooth muscle cells, pericytes, and other less well defined "interstitial cells".[7]

Function

Collagen fibers are arranged in a spiral surrounding airways, vessels and alveoli and in a continuous network extending from the hilum to the pleura. Because of their resistance to stretching and little, if any, elasticity, collagen fibers regulate compressibility and distensibility of bronchopulmonary structures, but play no role in lung elastic recoil. The relative role of elastin and collagen is nicely illustrated in Fig. 1, taken from Turino.[9] In experimental animals, elastase induces loss and disarray of elastin fibers which are believed to be the cause of the loss of aveolar attachments, the increase in the distensibility of the lung, and the decrease in the ability of the lung to recoil during expiration. Under these circumstances the alveoli become overdistended and break down causing the irreversible enlargement of airspaces that characterizes emphysema. By contrast collagenase, which mainly attacks collagen, does not modify the tissue recoil characteristics and thus does not cause hyperinflation, but only alters the pressure-curve at high lung volumes due to a reduction in tissue restraint provided by collagen (Fig. 1). In other words, experimental emphysema can be induced by elastase but not by collagenase.

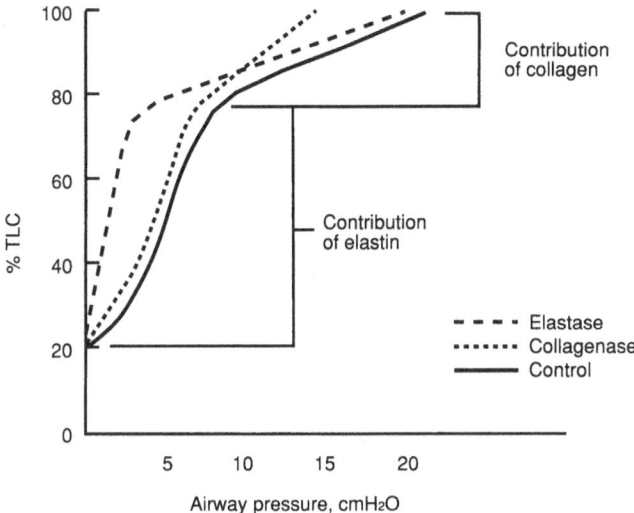

Fig. 1. A composite diagram of the effects of intratracheal elastase on the pressure-volume behaviour of whole lung as compared to collagenase. After elastase, there is a loss of recoil with a shift of the pressure-volume relationship to the left over the lower 80% of volume. There is still a relative flattening of the curve at the higher lung volumes after elastase, which can be attributed to the restraining effect of collagen. After collagenase, the slope increased at the higher lung volumes. (Reprinted from Turino[9] with permission).

Elastic fibers appear wavy when unstretched and lose their waviness when stretched. This change is attributed to a more parallel arrangement of the elastin polypeptide chains within the fibers, and provides the molecular structure underlying the characteristic ability of lung tissue to return to the original form once the stretching force is removed. Glycosaminoglycans and fibronectin serve mainly as a "glue" for collagen fibers and elastin, but they also have an important role in regulating cell movement (e.g. chemotaxis of mesenchymal cells), cell attachment, cell proliferation and tissue differentiation.

In addition to providing structural support for the cells and influencing the mechanical properties of the lung, the pulmonary interstitium, the basement membranes and the pleura have an important role in the modulation of the passage of fluids, solutes and gases, and contribute to the defence of the lung by creating a mechanical barrier to exogenous insults.[8]

The mesenchymal cells may greatly influence both the mechanical properties of the lung and the biochemical composition of the fibroskeleton, being the most important producers of collagen, proteoglycans and fibronectin. Their turnover is very slow, but can be accelerated in diseases such as fibrosis.

Biosynthesis of Pulmonary Interstitium Components

Collagen I and III, glycosaminoglycans and fibronectin, are synthesized mainly by the mesenchymal cells, whereas elastic fibers and collagen IV may be synthesized also by smooth muscle, epithelial and endothelial cells. Epithelial and endothelial cells can produce type I but mainly type III collagen, in addition to glycosaminoglycans and fibronectin. Thus, these cells may contribute to the final structure not only of the basement membrane but also of the entire pulmonary interstitium. For instance, the normal ratio of collagen I/III is 2:1 in the normal lung; this represents the balance of production of different types of collagen by mesenchymal and epithelial and endothelial cells.

This ratio increases in pulmonary interstitial fibrosis probably because of the increase in mesenchymal cells, and this increased proportion of the rigid type I collagen contributes to the increased stiffness of the fibrotic lung. The physiologic turnover of pulmonary interstitium components and mesenchymal cells after complete maturation is extremely slow.[9]

Because of the interest of elastin turnover for the pathogenesis of emphysema, we will describe the structure and turnover of elastin in more detail. Elastin derives from its precursor tropoelastin, a single polypeptide chain of about 800 residues, with a molecular weight of approximately 68 kd. Characteristically, elastin is made mainly by non polar amino acid, and has cross linking aminoacids that allow the linkage between the soluble tropoelastin and the insoluble fibers.

The step of elastin synthesis most relevant to the pathogenesis of emphysema is the extracellular cross linking of soluble elastin by oxidative deamination of lysine residues to form the amino acids desmosine and isodesmosine, a biochemical step necessary to develop mature elastic fibers extracellularly.

After lung injury, the turnover of elastin is greatly increased, with increased degradation followed by active resynthesis mainly dependent on lysil-oxidase activity. Genetically deficient animals or reduction of lysil-oxidase activity (e.g. by tobacco smoke) may be involved in the pathogenesis of pulmonary emphysema because of the inability to repair damaged elastic fibers.

Degradation of Pulmonary Interstitium Components

The composition of the pulmonary fibroelastic skeleton reflects the equilibrium between synthesis and degradation of the cellular and matrix constituents. Proteases are responsible for the degradation of connective tissue proteins: thus collagenase cleaves mainly collagen, elastase cleaves mainly elastin, and cathepsin B, a cysteine protease present in many cells including fibroblasts and alveolar macrophages, degrades soluble collagen, collagen fragments, elastin and glycoproteins.

In the normal lung collagenase is produced by fibroblasts and, in lower amounts, by macrophages. The only known source of elastase in normal condition is the alveolar macrophage, but the tiny amount produced justifies the extremely slow turnover of elastin.

During an inflammatory process, the main sources of collagenase and elastase are neutrophils and eosinophils. In fact, these two enzymes, and particularly elastase, are really broad proteases able to cleave most connective tissue proteins.

Oxygen radicals can degrade connective tissue, but it is not known whether they participate in the physiologic turnover of lung interstitium and, more importantly, in its degradation during inflammatory reactions.

Pathogenesis of Emphysema

As previously mentioned, emphysema is a "condition of the lung characterized by enlargement of airspaces distal to the terminal bronchioles accompanied by destruction of their walls, and without obvious fibrosis".

Thus, in this disease the interstitium is affected, and although all connective tissue components may be affected, the elastic fibers are primarily damaged.

A large body of circumstantial evidence has accumulated in the last 20 years suggesting that pulmonary emphysema may be due to unrestrained proteolytic activity in lung connective tissue causing extensive damage to the elastic fibers.

Alpha-1-antiproteinase (α-1-PI) is a 52 kd protein synthesized by the liver, capable of inhibiting several types of proteolytic enzymes. It is present in the serum (130-200 mg/100 ml) and in bronchoalveolar lavage fluid. Because of its activity, it is believed to represent a major defense mechanism against proteolytic attack in several organs, and in fact it increases during different inflammatory stimuli.

Homozygous deficiency of α-1-PI is associated with premature development of severe emphysema,[10] and this observation represents the cornerstone of the hypothesis that emphysema results from an imbalance between proteases and antiproteases in the lungs. This hypothesis has been supported by several experimental observations, the most important being that instillation of human neutrophil elastases into the lung as well as inhibition of α-1-PI synthesis may cause emphysema or make the animal more susceptible to emphysema.[11]

In man, in addition to the genetic deficiency of α-1-PI, the increase of proteolytic activity may result from an increased amount of proteases carried by inflammatory cells, particularly neutrophils.[12-14] Exogenous inflammatory stimuli (e.g. cigarette smoke, gaseous pollutants, endotoxins, etc.) may increase the proteolytic activity by attracting neutrophils into the lung and stimulating these cells to release their proteolytic burden. However, some of these stimuli (e.g. cigarette smoke) may also inhibit or inactivate the α-1-antiproteinase.

Oxygen radicals generated by inflammatory stimuli such as cigarette smoke may be responsible for reversible inhibition or inactivation of α-1-PI. [12-13]

An alternative mechanism that may be important in the pathogenesis of emphysema and that not been explored in due detail is represented by impaired elastin resynthesis after injury. The instillation of protease in the lungs of experimental animals is followed by an increase in both elastin fragments and newly synthesized elastin, in addition to an increased activity of lysil-oxidase, the enzyme that cross-links soluble elastin in the extracellular space, a biochemical step necessary to develop mature elastic fibers extracellularly.

Stimuli capable of inhibiting lysil-oxidase activity (e.g. cigarette smoke, lathyritic agents) or of interfering with protein assembly (e.g. elastase damaging microfibrillar components of elastic fibers) may interfere with the mechanisms of the lung and enhance the development of emphysema.

References

1. Macklem P.T.: Hyperinflation. Editorial. Am. Rev. Respir. Dis. 1984; 129: 1-2.
2. Thurlbeck W.M. :Overview of the pathology of pulmonary emphysema in the human. Clin. Chest Med. 1983; 4: 337-350.
3. Saetta M., Shiner R.J., Angus G.E. et al.: Destructive index: a measurement of lung parenchymal destruction in smokers. Am. Rev. Respir. Dis. 1985; 131: 764-769.
4. Saetta M., Ghezzo H., Kim W.D. et al.: Loss of alveolar attachments in smokers. A morphologic correlate of lung function impairment. Am. Rev. Respir. Dis. 1985; 132: 894-900.
5. Snider G.L, Kleinerman J., Thurlbeck W.M., Begali Z.H.: The definition of emphysema. Report of a National Heart, Lung and Blood Institute, Division of Lung Diseases Workshop. Am. Rev. Respir. Dis. 1985; 132: 182-185.
6. Staub N.C., Albertine K.H.: The structure of the lungs relative to their principal function. In: Murray J.F. Nadel J.A. (Eds). *Textbook of Respiratory Medicine*. Philadelphia, Saunders WB Publisher, 1988, pp. 12-36.
7. Cantor J.E., Turino G.M.: The pulmonary fibroelastic skeleton: a functional perspective. Chapter 73. In: Fishman A.P. (Ed) *Pulmonary diseases and disorders*. Second edition. New York, McGraw-Hill Book Publisher, 1988. pp. 1201-1208
8. Crystal R.G., Ferrans V.J., Turino G.M.: Reaction of the interstitial space to injury. Chapter 48. In: Fishman A.P. (Ed). *Pulmonary diseases and disorders*. Second edition. New York, McGraw-Hill Book Publisher, 1988. pp. 711-738.
9. Turino G.M.: The lung parenchyma - a dynamic matrix. J. Burns Amberson Lecture. Am. Rev. Respir. Dis. 1985; 132: 1324-1334.
10. Janoff A.: Elastase and emphysema. Am. Rev. Respir. Dis. 1985; 132: 417-433.
11. Snider G.L., Lucey E.C., Stone P.J.: Animal models of emphysema. Am. Rev. Respir. Dis. 1986; 133: 149-169.
12. Snider G.L.: Chronic bronchitis and emphysema. Chapter 44. In: Murray J.F. and Nadel J.A. (Eds). *Textbook of Respiratory Medicine*. Philadelphia, Saunders WB Publisher, 1988, 1069-1106.
13. Senior R.M., Kuhn C. III.: The pathogenesis of emphysema. Chapter 74. In: Fishman A.P. (Ed). *Pulmonary diseases and disorders*. Second edition. New York, McGraw-Hill Book Publisher, 1988. pp. 1209-1218.
14. Bieth J.G.: L'élastase leucocytaire humaine. Path. Biol. 1988; 36: 9: 1108-1111.

3. Hyperinflation and Trapped Gas in Chronic Airflow Limitation

S.M. Pritchard, J.C. Waterhouse, P. Howard
Respiratory Function Unit, Department of Medicine, Royal Hallamshire Hospital, University of Sheffield, England

Introduction

Lung volume is determined in part by the elastic forces of the resting thoracic cage. At functional residual capacity (FRC), the elastic recoil of the respiratory system is zero. The lung expands from this point by the application of inspiratory muscle effort along the linear portion of the thoracic compliance curve and in doing so maintains ventilation at the most cost effective point in terms of energy expenditure. In chronic airflow limitation resistance develops in the airways, particularly small airways, and the elastic recoil of the lung is diminished by destruction of elastin in alveolar walls. The resting volume at end expiration is associated with an increased alveolar pressure termed intrinsic PEEP (positive end expiratory pressure). More inspiratory work is required to overcome this pressure and expand the lung which by then may have been pushed into the non-linear part of the compliance curve. As the time constant of alveolar emptying is lengthened increased rate of breathing will further increase lung volume. These mechanical changes are likely to affect ventilation during acute exacerbations and on exercise, but will also be present to some extent in convalescence.

Bronchodilator drugs reduce airways resistance and allow end expiratory lung volume to fall by the release of so-called trapped gas. This is probably the most important consequence of bronchodilator therapy in chronic airflow limitation. Lung volume might be very sensitive to changes in small airway resistance and be directly related to symptoms of dyspnoea.

Uniphyllin Continus (Napp Laboratories Ltd) is a theophylline in a particularly stable slow release preparation. A remarkably even serum level is maintained throughout the 24h period which can be adjusted using the formula of Chrystyn et al.[1] In stable patients with chronic airflow limitation and little reversibility to β_2-agonists the effect of a daily dose of Uniphyllin was investigated to determine whether any improvement of breathlessness could be related to changes in spirometric values, residual volume or trapped gas volume.

Methods

Patients with advanced chronic obstructive airways disease were invited to take part in the study. They were to be in a stable phase, free of exacerbation or infection. Fig. 1 shows a flow diagram of the study. The patients were to receive, during two consecutive three week periods, active drug and placebo assigned at random in a double blind fashion. There were three attendances, a baseline and after each treatment period.

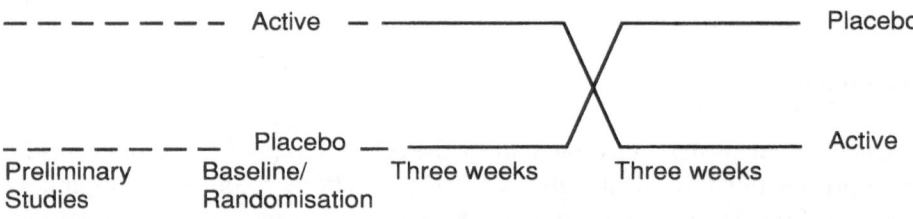

Fig. 1. Flow diagram of the study.

As a preliminary study the patients had a dose regime assessment. The dose of Uniphyllin necessary to give a steady state serum theophylline of 14 mg/l was calculated using the basic prediction nomogram of Chrystyn et al.[1] The patients attended the laboratory and attempted all the tests they were asked to perform.

They gave informed consent for the study and were issued with half the daily dose of Uniphyllin for four days and the full dose for the subsequent three days. On the seventh day a blood sample was taken and the patient subsequently had no theophylline therapy until the main phase of the study. The serum theophylline level was then used in the Bayesian analysis section of the prediction formula and a revised dosage was calculated if necessary.

Patient Selection Criteria

The patients were to have chronic obstructive airways disease, be less than 70 years of age, have an FEV/FVC ratio less than 70%, and a FEV_1 of less than 65% predicted. Reversibility to an inhaled selective β_2- agonist, terbutaline, was measured after a dose of 5 mg and had to be less than 15%. Patients were also to have no clinical evidence of heart disease.

Measurements Made

A history was taken and physical examination made at the preliminary study. At each of the three study visits patients performed spirometric tests of FEV_1 and FVC (Micromedical Instruments). Airway resistance (Raw) and thoracic gas volume were measured by the P.K. Morgan body plethysmograph[2-3] and lung volumes by the helium dilution technique (Gould Godart), and breathlessness (VAB) at both beginning and end. Arterial blood gas analysis followed with a venous blood sample for theophylline level.[4-6]

Results

15 patients were studied (10 male) with a mean age ($\pm$ SD) of 66.2 years[7]. FEV_1 at entry was 1.25 (0.56) litres, FVC 2.41 (0.9) litres and residual volume as a percentage of total lung capacity measured by the helium dilution technique (RV/TLC He) was 54.[8-9]

Table I shows mean values at the three visits. Changes between active therapy and placebo are in the last column. There was a small improvement in FEV_1 and FVC during the period of active therapy. The values are commensurate with expectations considering patient selection criteria of less than 15% reversibility to terbutaline. Specific conductance improved marginally.

Trapped gas volume (TGV), calculated as the difference between FRC measured in the plethysmograph (FRC box) and FRC by helium dilution (FRC He), fell by 0.3 litres on active therapy and only 0.1 litre during the placebo period, a 27% fall. There were small changes in residual volume (RV) by both methods. Walking distance and breathlessness improved but by a small, non-significant amount. Arterial blood gases improved slightly.

In Table II, changes in breathlessness are related to changes in the various respiratory function tests between active and placebo treatment.

Absolute change in breathlessness correlates highly with change in trapped gas volume and a little less with change in FRC measured by either plethysmography or helium dilution. Improvement of breathlessness is not quite significantly related to change in RV (He), FEV_1 or FVC. Considering the fact that the patients were selected because of their limited reversibility to a β_2-agonist, the change in trapped gas volume is striking and illustrates how changes in trapped gas volumes may occur in the face of trivial changes in FEV_1 or FVC during inhaled bronchodilator therapy. In Fig. 2 the change in breathlessness is plotted against the change in trapped gas volume.

Table I. Mean values ($\pm$ S.D.) for measurements made at each study visit and the percentage difference between placebo (P) and active treatment (A) and its significance. VAB was measured in mm using a 100 mm visual analogue scale with "no breathlessness" on the left and "severe breathlessness" on the right.

FEV_1 forced expiratory volume in one second, FVC: forced vital capacity, sGaw: specific airways conductance, RV: residual volume: TLC: total lung capacity, PaO_2 and $PaCO_2$: arterial blood gases, FRC: functional residual capacity, L: litre

		Baseline	Placebo (P)	Active (A)	% change (P-A)	p value
FEV_1	L	1.16 (0.56)	1.12 (0.43)	1.26 (0.61)	+11	< 0.05
FVC	L	2.41 (0.90)	2.33 (0.82)	2.66 (0.98)	+14	< 0.01
sGaw	ml/kPa/L	0.084 (0.046)	0.073 (0.028)	0.089 (0.033)	+20	< 0.05
PaO_2	kPa	9.9 (1.4)	9.9 (1.3)	10.4 (1.3)	+5	ns
$PaCO_2$	kPa	5.3 (0.5)	5.3 (0.4)	5.2 (0.5)	-2	ns
RV (He)	L	3.41 (1.4)	3.53 (1.3)	3.33 (1.3)	-6	=0.05
RV/TLC (He) %		54 (9)	56 (7)	52 (9)	-8	< 0.01
FRC (He)	L	4.28 (1.57)	4.21 (1.44)	4.11 (1.43)	-2	ns
FRC (box)	L	5.23 (1.62)	5.28 (1.76)	3.91 (1.56)	-7	< 0.05
Trapped gas	L	0.95 (0.42)	1.11 (0.90)	0.81 (0.51)	-27	ns
6 min walk	m	369 (110)	389 (80)	397 (97)	+2	ns
VAB	mm	66 (23)	65 (23)	62 (25)	-6	ns

Table II. Correlation coefficients, and their significance, between breathlessness and lung function tests. Symbols as in Table I

Parameter	r value	p value
Trapped gas	0.81	<0.01
FRC (box)	0.66	<0.01
FRC (He)	0.50	<0.05
RV (He)	- 0.40	ns
RV/TLC (He)	0.20	ns
FEV$_1$	0.45	0.07
FVC	0.46	0.07
sGaw	0.31	ns

Although the majority improved, a small number of patients deteriorated. The change in trapped gas volume was highly related to breathlessness whether the patients improved or deteriorated.

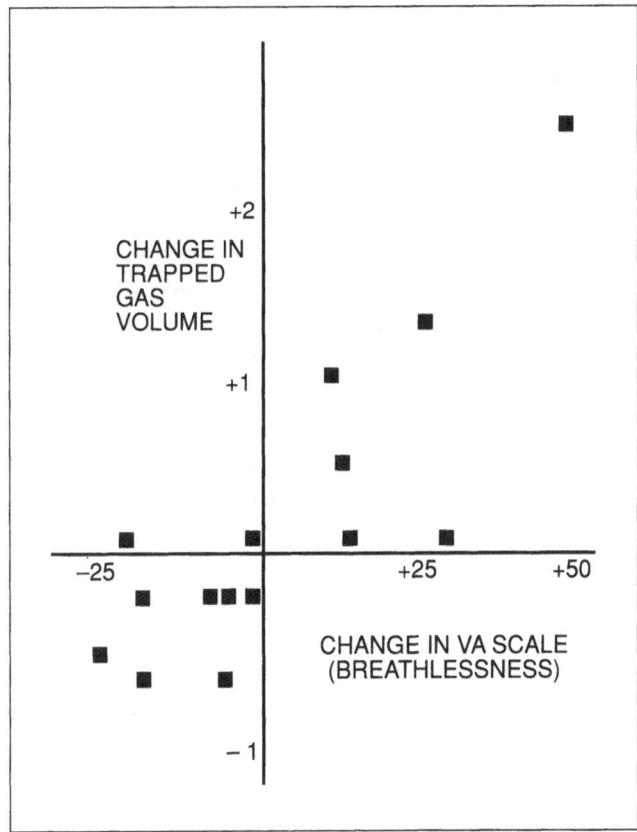

Fig. 2. Change in breathlessness in absolute terms compared to change in trapped gas volume in litres.

Discussion

Increasing hyperinflation of the lungs has long been associated with increased breathlessness. Rising functional residual capacity is an important component of hyperinflation. Much of the increased functional residual capacity is recorded as residual volume using the helium dilution technique. The same values are generally much higher when measured in the body plethysmograph. This difference is interpreted as being a large poorly ventilated volume not penetrated by the helium mixing technique. The gas cannot be completely separated from communication with the airway as absorption collapse of lung would follow. In the current study trapped gas volume has been calculated at FRC. These values may be different from other values in the literature where TLC has been employed.

Bronchodilator therapy may give benefit through reduction of trapped gas volume in patients with little or no spirometric reversibility. Changes in breathlessness were associated more strongly with changes in trapped gas volume than with changes in spirometry and this was whether the patients improved or, in one or two cases, deteriorated. In chronic airflow limitation the nature of the sensation of breathlessness is not understood. It is known that changes in lung volume induced by postural changes can sharply increase breathlessness.[6-7] According to theories of length tension inappropriateness[8] reduction of trapped gas volume would be expected to be associated with reduced swings of pleural pressure during tidal breathing as a result of (a) the lung moving on to a more linear part of its compliance curve and (b) reduction in airway resistance. The imbalance between desired ventilatory volume and muscle tension required to produce it would diminish. Bronchodilator therapy using Uniphyllin to produce a steady 24h blood level could plausibly improve breathlessness in this manner.

A moderate, inverse correlation was observed between TGV and RV(He). This supports the hypothesis that bronchodilation increases the volume of lung available for gas mixing, thereby reducing trapped gas volume. However, trapped gas volume may be in part due to an artefact described by Rodenstein et al.[9-10] They suggest that internal air flows set up by the movements of compliant extrathoracic airways, acting across an increased airways resistance, lead to an underestimation of intrinsic PEEP and thus thoracic gas volume. Similarly Brown showed that air in the gut can theoretically lead to a wrong estimation of absolute thoracic gas volume.[11] Thoracic gas volume measured in the body plethysmograph is likely to consist of a number of components but nevertheless its major change is associated with a change in intrinsic airways resistance. When bronchoconstriction is induced experimentally by the inhalation of carbachol, airways resistance and FRC box increase in both normal subjects and patients with COAD and return to resting values after sympathomimetic reversal.[12]

The disparity between changes in lung volume and spirometry and falls in thoracic gas volume are particularly evident during exacerbations of bronchitis or bronchial asthma. Corbeel[13] observed that the decrease in thoracic gas volume during treatment of bronchitic patients was almost entirely accounted for by a fall in FRC box. Similarly Woolcock et al.[14] found FRC box doubled during severe exacerbations of asthma only to fall again during recovery. Peake et al.[15] compared recovery from exacerbations of both asthma and bronchitis. They observed FRC box fall but FRC He rose during recovery from exacerbations of bronchitis. FEV_1 and airway conductance changed little despite marked clinical improvement. In asthmatics FEV_1 increased by a mean of 152%. Nebulized terbutaline given a week after the acute episode increased the asthmatics' FEV_1 by 57% with no change in thoracic gas volume, suggesting a different mechanism is responsible for the acute response to inhaled bronchodilators, as opposed to the more complex changes occurring during recovery from acute exacerbations. In airflow limited subjects response to bronchodilator therapy may therefore need to be measured by techniques other than routine spirometry.

In conclusion, hyperinflation is important to the sensation of breathlessness in patients with fixed airways obstruction.

Improvements in trapped gas can improve breathlessness when spirometric benefits seem few. In acute exacerbations benefits of reducing airway resistance are more complex. Long term theophylline therapy can benefit patients with chronic airflow limitation.

References

1. Chrystyn H., Mulley B.A., Peake M.D.: Precise individualisation of theophylline dosage using a nomogram and Bayesian analysis and dependence of accuracy on preparation used. In: M. Turner-Warwick J. Levy (Eds.) *New prospectives in theophylline therapy*, International Congress and Symposium series n° 78, Royal Society of Medicine 1984; 117-128

2. Bedell G.N., Marshall R., Dubois A.B., Comroe J.H.: Plethysmographic determination of the volume of gas trapped in the lungs. J. Clin. Invest. 1956; 35: 664-670

3. Du Bois A.B., Botelho S.Y., Bedell G.N., Marshall R., Comroe J.H.: A rapid plethysmographic method for measuring thoracic gas volume: a comparison with a nitrogen washout method for measuring functional residual capacity in normal subjects. J. Clin. Invest. 1956; 35: 322-326

4. McGavin C.R., Gupta S.P., McHardy G.J.R.: Twelve minute walking test for assessing disability in chronic bronchitis. Br. Med. J. 1976; 1; 822-823

5. Butland R.J.A., Pang J., Gross E.R., Woodcock A.A., Geddes D.M. : Two, six and twelve minute walking tests in respiratory disease. Br. Med . J. 1982; 284: 1607-1608

6. Sharp J.T., Drutz W.S., Moisan T., Foster J., Macnach W.: Postural relief of dyspnoea in severe obstructive pulmonary disease. Am. Rev. Respir. Dis. 1980; 122: 201-211

7. Druz W.S., Sharp J.T.: Electrical and mechanical activity of the diaphragm accompanying body position in severe chronic obstructive pulmonary disease. Am. Rev. Respir. Dis. 1982; 125: 275-281

8. Campbell E.J.M., Howell J.B.L.: The sensation of breathlessness. Br. Med. Bull. 1963; 19: 36-40

9. Rodenstein D.O., Stanescu D.C., Francis C.: Demonstration of failure of body plethysmography in airway obstruction. J. Appl. Physiol. 1982; 52: 949-954

10. Stanescu D.C., Rodenstein D.O., Cauberghs M., Van de Woestyne K.P.: Failure of body plethysmography in bronchial asthma. J. Appl. Physiol 1982; 52: 939-948

11. Brown R., Hoppin F.G., Ingram R.H., Saunders N.A., McFadden E.R.: Influence of abdominal gas on the Boyle's Law determination of thoracic gas volume. J. Appl. Physiol. 1978; 44: 469-473

12. Lovejoy F.W., Constantine H., Flatley J., Kaltreider N., Dautrebande L.: Measurement of gas trapped in the lungs during acute changes in airway resistance in normal subjects and in patients with chronic pulmonary disease. Am. J. Med. 1961; 30: 884-892

13. Corbeel L.J.: Comparison between measurement of functional residual capacity and thoracic gas volume in chronic obstructive pulmonary disease. Progr. Resp. Res. 1968; 4: 194-204

14. Woolcock A.J., Rebuck A.S., Cade J.F. Read J.: Lung volumes changes in asthma measured concurrently by two methods. Am. Rev. Respir. Dis. 1971; 104: 703-709

15. Peake M.D., Freestone S., Howard P.: Changes in trapped gas volume and other tests of airflow obstruction in exacerbations of chronic obstructive airways disease (COAD) and asthma. Eur. J. Resp. Dis. 1981; 62 (Suppl 113): 170-171

4. Effects of Airways Resistance on Lung Inflation

C. TARDIF[1], G. BONMARCHAND[1], B. ORCEL[2], J.P. DERENNE[2]
1. Respiratory Physiopathology Group, Ch. Nicolle Hospital, Rouen, France
2. Pneumology Service, Saint-Antoine Hospital, Paris, France

Chronic obstructive pulmonary diseases (COPD), i.e. chronic bronchitis, and emphysema, are characterized by anatomical and functional damage of the airways.

The first implications of these abnormalities can be observed when examining the flow volume loop. The expiratory part becomes concave, and the expiratory flows are reduced, mainly when measured at low lung volume, i.e. when they are independent of the expiratory effort. In severe COPD, forced expiratory flow is smaller than that measured during tidal breathing. This is due to an increased collapsibility of the airways, and to airway closure when intrathoracic pressure becomes positive. It follows that when the severe COPD patients use their expiratory muscles, they cannot increase, and often they may actually decrease expiratory flow instead of increasing it. Thus, when an increase in ventilation is needed, COPD patients cannot use the "expiratory" pathways, i.e. increase expiratory flow, or decrease lung volume but they have to develop a different strategy. In fact, there remain the "inspiratory" pathways: they can increase the amplitude of the negative inspiratory thoracic pressure, and they can breathe at an increased volume, which decreases airways resistance.

A number of observations indicate that the patients use both compensatory mechanisms. During tidal breathing in COPD patients, minute ventilation remains identical or slightly higher than in normal subjects, even during acute respiratory failure (ARF). These findings do not support the former belief that COPD patients undergoing ARF had global pulmonary hypoventilation. Since minute ventilation V is the product of tidal volume VT and respiratory frequency f, this maintained or increased V is obtained with an increased breathing frequency and a decreased VT.[1]

The inspiratory flow compensation can be deduced from the analysis of the breathing pattern:

$$V = VT/TI \times TI/TTOT = VT/TE \times TE/TTOT$$

where TI, TE and TTOT are inspiratory and expiratory times, and total breath duration respectively. If VT/TE decreases, TE/TTOT must increase. Since TTOT = TI + TE, the duty cycle TI/TOT decreases, and therefore mean inspiratory flow VT/TI must increase. In fact, the volumes measured every 100 msec at the very beginning of inspiration are greater than those of normal subjects, and of stable COPD patients.[2] However, this increased inspiratory flow may lead to inspiratory muscle fatigue.[3]

The second compensatory mechanism leads to hyperinflation, a common fact demonstrated by clinical and X-ray observations and by measurements of functional residual capacity. Examination of pressure volume curves reveals the main advantage of this new situation: elastic recoil pressure increases. Furthermore, airway resistance decreases when lung volume increases. Therefore the expiratory resistive work decreases. The new configuration of the expiratory muscles may also provide a better mechanical advantage.

However, hyperinflation has important consequences for inspiratory muscle action and coordination. The vertical displacement of the diaphragm may be reduced, the costal insertions may no longer be vertical, and the surface of the costal apposed portion of the diaphragm is reduced. This explains why during inspiration the lower part of the rib cage may move inward (Hoover's sign). Diaphragmatic fiber length decreases, which diminishes the efficacy of their contraction, i.e. less pressure is generated for a given stimulus. Actually, gastric pressure swing is small, and even often absent during inspiration.[4] Thus, the contractile activity of the diaphragm seems to be more used to prevent transmission of negative thoracic pressure to the abdomen, and to prevent the abdomen from being sucked into the thorax during inspiration. This implies that the diaphragm may rather be used as a fixator of the thoracic basis than as a major agonist of inspiration.

Therefore, the main inspiratory agonistic work must be performed by the intercostal and inspiratory accessory muscles. The scalene and sternomastoid muscles contract vigorously during inspiration in severe COPD patients and the "respiratory pulse", i.e. their contraction during tidal breathing, can be observed and palpated.

The modifications of the respective roles of the inspiratory muscles induce modifications of the biochemical and energetic structure of the muscles fibers. There is a decrease in type I and II diaphragmatic fiber diameters, proportional to the decrease in forced expiratory volume in one second FEV_1 and vital capacity VC.[5,6] There are decreased maximal enzymatic activities of hexokinase and lactic dehydrogenase, i.e. of key enzymes controlling metabolic anaerobic activities, in proportion to the decrease in FEV_1 and VC and to the degree of hyperinflation.[7] These modifications may be responsible for a diminished resistance to fatigue. Furthermore, Sanchez et al.[8] found compensatory increased maximum metabolic activities of hexokinase, 3-hydroxyacylCoA dehydrogenase, and citrate synthase in the internal and external intercostals of COPD patients, proportional to the degree of airways obstruction and pulmonary overinflation. This indicates greater use of these muscles by COPD patients, and supports the hypothesis that both intercostal muscles are essentially inspiratory.

In the acute phase of respiratory failure, a common observation is the contraction of the abdominal expiratory muscles during expiration. This activity cannot induce an increase in expiratory flow, which is mechanically limited.

However, it may permit the storage of elastic energy during expiration and its release at the beginning of inspiration could promote inspiration without generating a transdiaphragmatic pressure (Pdi), thus protecting the diaphragm, a phenomenon which has been documented in normal man.[9]

Shortening of diaphragmatic fibers leads to a diminished contractile efficiency. The diaphragm and accessory muscles must contract more vigorously than those of normal subjects in order to produce the same flow and volume. The increase in inspiratory flow is associated with an increased Pdi.[2] Inspiratory muscle fatigue occurs when the tension generated is greater than 40 percent of the maximal tension.[10]

In these conditions, the energy consumption in the muscle is greater than the blood supply of energy. The diaphragm may reach its fatigue threshold, which has been demonstrated to occur when the tension time index (TTdi = TI/TTOT x Pdi/Pdi max) is higher than 15[3], where Pdi is the mean transdiaphragmatic pressure during inspiration and Pdi max is maximum transdiaphragmatic pressure.

Central repiratory drive must be increased in order to produce normal or increased ventilation with a ventilatory apparatus characterized by an increased mechanical hindrance and a reduced mechanical efficiency of the diaphgm. In fact, the occlusion pressure P0.1, an index of central inspiratory activity,[11] is markedly increased during acute respiratory failure.[1-2]

Nevertheless, in spite of this increased inspiratory drive, the blood gases of severe COPD patients are grossly abnormal, and these abnormalities are aggravated during ARF.

Moreover, when additional oxygen is administered, there is an increased hypercapnia. This was classically interpreted as the consequence of a decreased sensitivity of the respiratory centers: because of chronic hypercapnia, CO_2 drive was supposedly blunted, and the increased hypercapnia was presumably the consequence of the removal of the hypoxic drive by the administration of oxygen. This classical theory implied that the hypoxic drive to breathe had a prominent role in the generation of respiratory activity in ARF. However, a number of data have contradicted this concept. Hypercapnia may be determined by metabolic factors, particularly metabolic alkalosis.[12] Pure oxygen inhalation induces very little decrease in ventilation in COPD patients with ARF, and there is no correlation between changes in $PaCO_2$ and ventilation.[13] After peripheral chemodenervation in 27 COPD patients followed for six years, no excess mortality could be detected.[14] This suggests that the hypoxic drive to breathe may not be of major importance in the vital compensatory mechanisms in COPD patients. Moreover, we recently tested the effect of rising CO_2 concentrations on P 0.1 and ventilation in 37 COPD patients and in 24 normal subjects. The patients were undergoing ARF, and had been mechanically ventilated. There was a marked decrease in the ventilatory response of the patients as compared to the normals (mean slope 0.18 versus 1.091/min/mmHg), while P 0.1 responses were nearly the same (0.28 versus 0.37 cmH$_2$0/mmHg)[15]. This implies that CO_2 sensitivity is not blunted in COPD patients with ARF, and may play an important role in respiratory control under those circumstances.

What the other drives to breathe are remains essentially unknown. However, we have some information about the role of the vagus nerves.

Metacholine provokes hyperinflation and increases airways resistance. This is associated with rapid and shallow breathing, increased P 0.1, and minute ventilation, and increased $PaCO_2$.[16]

By contrast, the instillation of xylocaine from the larynx to the subsegmental bronchi induced an increased VT, VT/TI, and TE.[17] These observations suggest that parasympathetic afferences are involved in the determination of breathing pattern and gas exchange in COPD patients, particularly during ARF. However, the changes observed are moderate. Moreover, the breathing pattern of respiratory patients was not dramatically modified after vagal blockade.[18]

This implies that there are other neural factors which interfere with respiratory drive in ARF, but there is little information available about their nature and importance.

References

1. Derenne J.Ph., Aubier M., Murciano D., Fournier M., Pariente R.: Contrôle de la respiration au cours des poussées d'insuffisance respiratoire aiguë des insuffisances respiratoires chroniques obstructives. Rev. Mal. Respir. 1977; 5: 714-716.
2. Derenne J.Ph., Fleury B., Pariente R.: Acute respiratory failure of chronic obstruction pulmonary disease. Am. Rev. Respir. Dis. 1988; 138: 1006-1033.
3. Bellemare F., Grassino A.: Effect of pressure and timing of contraction on human diaphragm fatigue. J. Appl. Physiol. 1982, 53, 1190-1195.
4. Murciano D., Aubier M., Bussi S., Derenne J.Ph., Pariente R., Milic-Emili J. : Comparison of

esophageal, tracheal, and mouth occlusion pressure in patients with chronic obstructive pulmonary disease during acute respiratory failure. Am. Rev. Respir. Dis. 1982; 126: 837-841.

5. Sanchez J., Derenne J.Ph., Debesse B., Riquet M., Monod H.: Typology of the respiratory muscles in normal men and in patients with moderate chronic respiratory diseases. Bull. Eur. Physiopathol. Respir. 1982; 18: 901-914.

6. Sanchez J., Medrano G., Debesse B., Riquet M., Derenne J.Ph.: Muscle fibre type in costal and crural diaphragm in normal men and in patients with moderate chronic respiratory disease. Bull. Eur. Physiopathol. Respir. 1985; 21: 351-356.

7. Sanchez J., Bastien C., Medrano G., Riquet M., Derenne J.Ph.: Metabolic enzymatic activities in the diaphragm of normal men and patients with moderate chronic obstructive pulmonary disease. Bull. Eur. Physiopathol. Respir., 1984; 20: 535-540.

8. Sanchez J., Brunet A., Medrano G., Debesse B., Derenne. J Ph.: Metabolic enzymatic activities in the intercostal and serratus muscles and in the latissimus dorsi of middle-aged normal men and patients with moderate obstructive pulmonary disease. Eur. Respir. J., 1988; 1: 376-383.

9. Grassino A.E., Derenne J.P., Almirall J., Milic-Emili J., Whitelaw W.A.: Configuration of the chest wall and occlusion pressures in awake man. J. Appl. Physiol., 1981; 33: 134-142.

10. Roussos C.S., Macklem P.T.: Diaphragmatic fatigue in man. J. Appl. Physiol. 1977; 43: 189-197.

11. Whitelaw W.A., Derenne J.P., Milic-Emili J.: Occlusion pressure as a measure of respiratory center output in conscious man. Respir. Physiol. 1975; 23: 181-199.

12. Mc Nicol M.N., Campbell E.J.M.: Severity of respiratory failure. Arterial blood gases in untreated patients. Lancet 1965; 1: 336-340.

13. Aubier M., Murciano D., Milic-Emili J., Touaty E., Daghfous J., Pariente R., Derenne J.Ph.: Effects of the administration of 0_2 on ventilation and blood gases, in patients with chronic obstructive pulmonary disease during acute respiratory failure. Am. Rev. Respir. Dis., 1980; 122: 747-754.

14. Vermeire P., De Backer W., Vanmaele R., Bal G., Van Kerckhoven W.: Carotid body resection in patients with severe chronic airflow limitation. Bull. Eur. Physiopathol. Respir. 1987; 23: (suppl. 11) 165S-166S.

15. Tardif C., Bonmarchand G., Gibon J.F., Leroy J., Pasquis P., Derenne J.Ph.: CO_2 sensitivity in patients with chronic obstructive pulmonary disease undergoing acute respiratory failure (submitted).

16. Oliven A., Cherniack N.S., Deal E.C.J, Kelsen S.G.: The effects of acute bronchoconstriction on respiratory activity in patients with chronic obstructive pulmonary disease. Am. Rev. Respir. Dis., 1985; 131: 236-241.

17. Murciano D., Aubier M., Viau F., Bussi S., Pariente R., Milic-Emili J., Derenne J.Ph.: Effects of airway anesthesia on pattern of breathing and blood gases in patients with chronic obstructive pulmonary disease during acute respiratory failure. Am. Rev. Resp. Dis., 1982; 126: 113-117.

18. Guz A., Noble M.I.M., Eisele J.M., Trenchard D.: The role of vagal inflation reflexes in man and other animals. In: R. Porter (Ed.) *Breathing : Hering Breuer centenary symposium*. London 1970, 17-40.

5. Alveolar Gas Mixing in Chronic Pulmonary Hyperinflation

F. CIBELLA, P. PIPITONE, C. MACALUSO, V. BELLIA, G. BONSIGNORE
Respiratory Physiopathology Institute, National Research Council, Pneumology Department, University of Palermo, Italy

Introduction

It is well known that in chronic pulmonary hyperinflation defects in intrapulmonary gas mixing and in $\dot{V}/\dot{Q}$ ratio are likely to occur.

These phenomena are usually evaluated by studying the convective gas distribution and the diffusive alveolar gas mixing; however, when performed by the multiple-breath nitrogen wash-out (m-b N_2 w-o) using the conventional methods based on the analysis of the expired gas concentrations, the relevant measurements are affected by high intra- and inter-individual variability. In fact, the results are critically dependent on variations of tidal volume, respiratory frequency and intrapulmonary gas volume.

Another method proposed for the evaluation of the impairment of gas exchange is represented by the non invasive analysis of expired carbon dioxide during quiet breathing in air (exp. CO_2 an.); however, the major limitation of this technique is the non simultaneous measurement of alveolar CO_2 and ventilation.[1]

To overcome these problems Cumming et al.[2] proposed to use measurement of gas volumes instead of concentrations in the analysis of m-b N_2 w-o and in exp.CO_2 an.; in our previous experiences this modification resulted in a marked improvement in the reproducibility of results.[3-4]

The simultaneous use of both methods allows the evaluation of intrapulmonary mixing of non diffusible gases (N_2) and CO_2 removal, as well as the measurement of serial (anatomical) and alveolar dead spaces for both gases. In this way it is possible to evaluate separately the defects in gas mixing due to impairment of diffusive and convective gas transport and to $\dot{V}/\dot{Q}$ mismatching.[5]

Accordingly, we applied this experimental approach to a sample of patients affected by chronic pulmonary hyperinflation with the following aims:

a) - to evaluate the degree of the impairment of convective and diffusive gas mixing which has been anticipated by Cumming on theoretical grounds;[5]

b) - to assess the alteration of CO_2 removal that is likely to result from the mismatching of the $\dot{V}/\dot{Q}$ ratio over the lung fields.

Samples and Methods

Normal Samples

To obtain the reference values for both the tests we evaluated two different samples of male healthy subjects (21 subjects for m-b N_2 w-o, including 10 smokers, and 17 for exp.CO_2 an., 7 smokers), respectively aged 29.0 years $\pm$ 3.15 SD and 30.5 years $\pm$ 3.2 SD, selected on the basis of the following characteristics:
- lack of family or personal history of respiratory diseases;
- no upper respiratory tract infections during the 15 days preceding the study;
- normal spirometric and plethysmographic indices.

Patient Sample

16 male subjects, affected by pulmonary hyperinflation (Table I), were submitted on the same day to a m-b N_2 w-o test and to exp.CO_2 an. during quiet breathing in air.

Table I. Functional characteristics of the patients (16 male patients)

Age	FEV$_1$(%pred)	RV/TLC%
65.1 ±7.6	42.6 ±19.6	68.1 ±11.8

Multiple-Breath Nitrogen Wash-out

During quiet breathing, from room air each subject was switched, through a three-way valve,[6] to an argon (79%)-oxygen (21%) mixture contained in a bag-in-box system equipped with a pneumotachograph (Fleisch No. 3); the subject was ventilated with this mixture until the end-tidal N_2 concentration dropped below 2%. The four inspiratory and expiratory gases (N_2, CO_2, O_2 and argon) were sampled at the mouth and analyzed by a mass spectrometer (Centronic MGA 200). The flow and gas concentration signals were recorded on tape (HP 3968A) and processed off line (Digital Vax 8200). Signals of flow and gas concentration for N_2, CO_2 and O_2 were sampled at 50 Hz; they were synchronized by considering the dynamic response of the system, measured by generating two synchronous signals (sudden changes in flow and in gas concentration) as an input of the system, and evaluating the time delay of the output. In addition, to avoid the variations in the time delay due to the breath by breath changes in expired gas composition, we applied the breath by breath correction for time delay using the adaptive synchronization suggested by Brunner et al.[7] Expiratory volumes of N_2, CO_2 and O_2 were calculated by means of the following equation:

$$Vgas = \int flow \times concentration = \int_{t=0}^{t=t} \frac{dV}{dt} \times F_E(t)dt$$

were dV/dt is the instantaneous flow rate and FE(t) is the instantaneous concentration of examined gas. The integration of the flow rate during a single expiration represents the gas volume in the expirate.[8] Serial dead space (VdS) was evaluated breath by breath by plotting the distribution of the total expired volumes (x-axis) versus the simultaneous cumulative volumes of expired N_2 (y-axis) (Fig. 1); the index was calculated within each breath as the intercept on the x-axis of a 2nd degree polynomial regression line relevant to the points sampled at 20 msec.

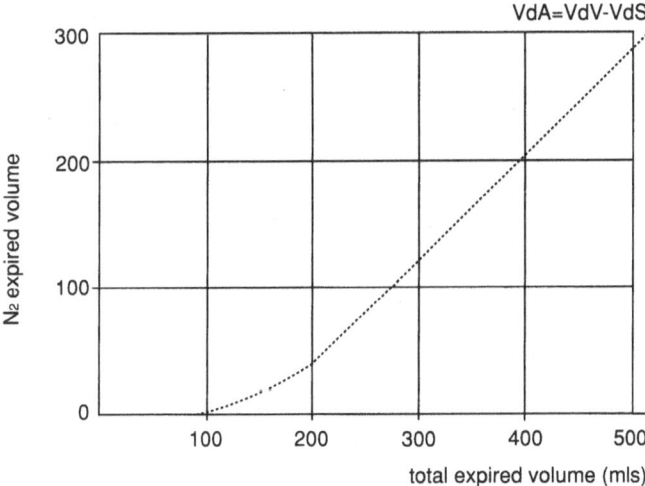

Fig. 1. Breath by breath serial dead space (VdS) is calculated as the intercept on the X-axis of a 2nd degree polynominal regression between the distributions of the total expired volumes and the cumulative volumes of expired N_2.

intervals (we retained the average VdS obtained from all the breaths examined during each wash-out test). Initial lung gas volume, i.e. the functional residual capacity (FRC), was calculated by summing all the N_2 expired volumes during the whole test and dividing by 0.79. The ventilation necessary for the loss of 90% of the intrapulmonary N_2 was calculated and normalized for the FRC (number of turnovers at 90% clearance): this index was compared with the ideal value, i.e. calculated for a lung with the same volume and ventilation, but without any dead space (Fig. 2).

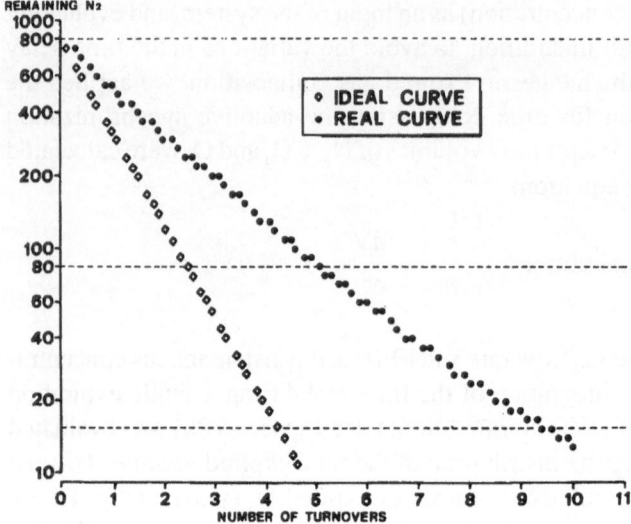

Fig. 2. Course of the ideal (diamonds) and real (circles) curves of N_2 wash-out: on the X-axis the ventilation (normalized for FRC and expressed as number of turnovers); on the Y-axis the logarithm of N_2 volume remaining in the lungs (normalized per liter of FRC).

On this basis the Total Ventilatory Efficiency (at 90% clearance) was calculated as the ratio of the observed number of turnovers to the ideal one, expressed as percentage. Its complement to 1, i.e. the ventilatory inefficiency, expressed as a fraction of the mean tidal volume (Vt), provides an estimate of the total ventilatory dead space (VdV). By subtracting the mean VdS from the VdV, the alveolar dead space for N_2 (VdAN$_2$) may be calculated.

The Alveolar Mixing Efficiency for N_2(N$_2$AME) can be evaluated by means of the following equation:

$$N_2 AME \% = \frac{Vt - VdS - VdAN_2}{Vt - VdS} \times 100$$

Now, the real alveolar ventilation (VA) can be computed:
VA = Vt - VdS - VdA

CO$_2$ Analysis During Air Breathing

By using the same experimental set-up, flow and gas concentrations (for N_2, CO_2 and O_2) were recorded and analyzed during quiet breathing in air. Expired CO_2 volume and VdS were evaluated breath by breath in the same way as for N_2.

The Alveolar Dead Space for CO_2 (VdACO$_2$) was calculated using the following equation:

$$VdACO_2 = Vt - VdSCO_2 - \frac{VCO_2 \times 100}{ETCO_2\%}$$

where: VCO_2 x $100/ETCO_2\%$ is equal to the CO_2 volume obtained when the CO_2 concentration is at its end-tidal value. In this case the measured CO_2 gas volume is relevant to the alveolar CO_2 volume.

Now the overall lung efficiency value for CO_2 removal (CO_2Alv.Eff.%) can be evaluated as follows:

$$CO_2 \text{ Alv. Eff. } \% = \frac{Vt - VdSCO_2 - VdACO_2}{Vt - VdSCO_2} \text{ x } 100$$

In this equation the mean values for Vt, $VdSCO_2$ and $VdACO_2$ are used.

Results

The mean values obtained in the patient sample from the m-b N_2 w-o test showed a N_2VdA much higher (more than twofold) with respect to the normal sample, as well as a lower VA; the N_2AME% was nearly halved with respect to reference subjects. For the exp.CO_2 an., the CO_2VdA was more than three times higher than in normal subjects with a significantly lower CO_2Alv.Eff.% (Table II). In Figs. 3 and 4 the partitioning of mean Vt (normalized for the same Vt value) is shown.

Table II. Results obtained from the nitrogen wash-out test and from the analysis of expired CO_2.

	Normal	Hyperinflated
N_2VdS	137.6 ± 15.5	169.2 ± 34.2
N_2VdA	92.3 ± 39.6	324.4 ± 105.9
N_2VA	423.3 ± 173.6	235.1 ± 170.3
N_2AME%	81.2 ± 4.6	39.1 ± 9.2
CO_2VdS	171.6 ± 21.6	170.4 ± 35.7
CO_2VdA	29.5 ± 8.5	115.7 ± 71.3
CO_2Alv.Eff.%	93.1 ± 2.6	81.5 ± 3.6

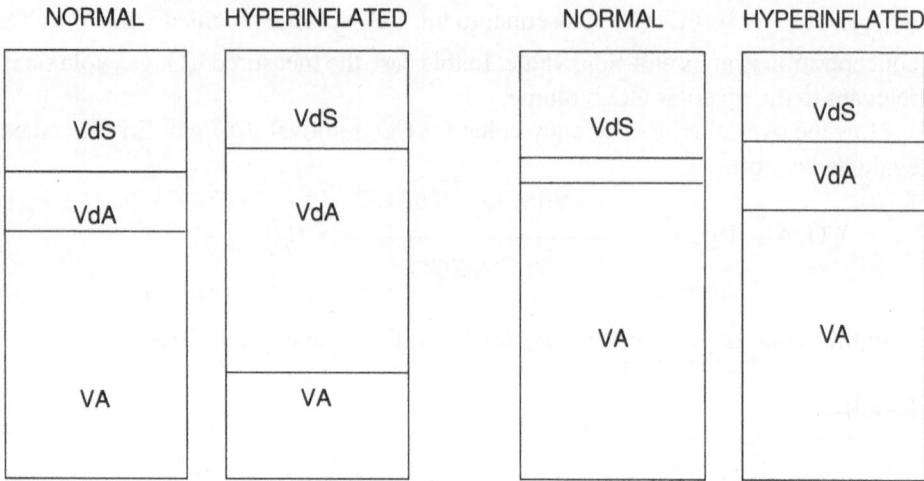

Fig. 3. Representation of the partitioning of mean tidal volume as obtained from the results of nitrogen wash-out.

Fig. 4. Representation of the partitioning of mean tidal volume as obtained from the results of expired CO_2 analysis during quiet breathing in air.

Discussion

The results of the present study illustrate the gross alteration in gas exchange occurring as a consequence of anatomical disorders manifested as chronic pulmonary hyperinflation: in fact, a marked increase in dead spaces and, consequently, a proportionate reduction in VA were observed. Although not easily comparable for methodological reasons, these results are in agreement with those previously reported in investigations based upon the m-b N_2 w-o evaluated following the conventional techniques of measurement of expired gas concentration and number of breaths.[9] However, the technique applied in the present study, providing reliable data relevant to the quantitation of dead spaces, allows the computation of the real VA.

The results obtained seem to confirm the theoretical assumption proposed by Cumming et al.[5] as regards the impairment of convective (airflow obstruction) and diffusive intrapulmonary gas mixing (destruction of lung alveolar walls with enlargement of areas in which diffusive mixing must occur) in pulmonary hyperinflation. In fact, the observed large increase in N_2VdA is compatible with the increase in distal air spaces, whereas the corresponding reduction in N_2AME% is an expression of the alteration in alveolar diffusive gas mixing. As regards the exp.CO_2 an. during quiet breathing, the results showed that in hyperinflated lungs the CO_2VdA was much higher than in normals. This can be considered an expression of the degree of impairment of CO_2 removal due to $\dot{V}/\dot{Q}$ mismatching. Similarly

CO_2 Alv.Eff.% was significantly lower in patients, even if the difference with respect to normal subjects was apparently smaller than for N_2AME%. This index is in fact representative of the incomplete adaptation of the pulmonary circulation to the defect of ventilation, so that the difference in the degree of impairment of CO_2Alv.Eff.% and of N_2AME% is not unexpected. The exp.CO_2 an., when performed with the methods outlined here, is a simple and non invasive technique that compares favourably with usual methods which are either more invasive[10] or, if non invasive, are affected by the methodological problems mentioned, i.e. the need for a previous and separate measurement of alveolar CO_2 (by rebreathing) before evaluating ventilation and expired gas concentrations (by a mixing chamber).[1]

In addition, in the latter method the measurement of anatomical dead space in patients with airway obstruction is prevented because of the use of expired gas concentrations. These methodological problems determined a large variability in the values obtained for the CO_2 Alv.Eff.% with unreliable estimates of this index (occasionally even higher than 100%).

In conclusion, the results of the application of the proposed method justify its use for a more thorough understanding of the pathophysiology of chronic pulmonary hyperinflation.

References

1. Bradley G.W., Winer J.: Alveolar efficiency for CO_2 in normal people and patients with chronic obstructive lung disease. Bull. Eur. Physiopathol. Resp. 1986; 22: 437-441.
2. Cumming G., Guyatt A.R.: Alveolar gas mixing efficiency in the human lung. Clinical Science, 1982; 62: 541-547.
3. Cibella F., Mangiacavallo A., Pipitone P., Macaluso C., Bonsignore G.: Evaluation of reproducibility of multibreath nitrogen washout test in normal subjects. In: Prog. Resp. Res. Vol. 21, Basel, Karger, 1986; 138-140.
4. Pipitone P., Cibella F., Macaluso C., Trizzino A., Corvo M., Bellia V., Bonsignore G.: Riproducibilita' di un test per lo studio del mixing dei gas diffusibili e sua applicazione clinica. Lotta contro la TBC e le malattie polmonari sociali. In press
5. Cumming G.: Gas mixing in disease. In: J.G. Scadding and G. Cumming (Eds.), *Respiratory Medicine*. London, W. Heinemann Medical Books Ltd., 1981; 688-697.
6. Lee K.D., Crisp H.A.: Solenoid operated valve box providing sudden undetectable changes of inspired gas. J. Appl. Physiol. 1974; 36: 765.
7. Brunner J.X., Wolff G., Cumming G., Langenstein H.: Accurate measurement of N_2 volumes during N_2 washout requires dynamic adjustment of delay time. J. Appl. Physiol. 1985; 59:1008-1012.
8. Cumming G., Jones J.G. :The construction and the repeatability of lung nitrogen clearance curves. Respir. Physiol. 1966; 1:238-248.
9. Paiva M., Yernault J.C., Martius D., Englert M.: Interpretation nouvelle des courbes de rincage d'azote. Bull. Eur. Physiopathol. Respir. 1974; 10:831-844.
10. Wagner P.D., Dantzker D.R., Duek R., Clausen J.L., West J.B.:Ventilation-perfusion inequality in chronic obstructive pulmonary disease. J. Clin. Invest. 1977; 59:203-216.

6. Acute Pulmonary Hyperinflation and Pulmonary Edema

D. DREYFUSS, G. SAUMON

Medical Resuscitation Service, Louis Mourier Hospital, Colombes, INSERM U82, Faculty of Xavier Bichat, Paris, France

Introduction

Mechanical ventilation is very useful in the treatment of pulmonary edema, whether of the hydrostatic or of the permeability type. Moreover, the use of positive end-expiratory pressure (PEEP) during mechanical ventilation consistently leads to better oxygenation.[1] Although it was initially hoped that PEEP might "push out" water from the lungs, such promise has not been fulfilled. Moreover, the use of PEEP has resulted in an increase in extravascular lung water content during experimental permeability edema.[2-3] This increase might have been the consequence of augmentation of fluid filtration through extraalveolar vessels, because of the resulting lung volume increase.[4-5] Whether such a change in lung water content may occur during PEEP ventilation of pulmonary edema in patients is difficult to prove; nevertheless, this possibility has been seriously considered.[6] Another cause of concern is the production of pulmonary edema as a consequence of lung acute hyperinflation. We will first present evidence that intermittent positive pressure ventilation (IPPV) with high inflation pressure (HIPPV) results in pulmonary edema; secondly we will discuss the mechanisms of HIPPV edema; and thirdly we propose to anbalyse the respective roles of airway pressure and lung volume in the production of edema, and how PEEP influences it.

Pulmonary Edema During HIPPV

Webb and Tierney[7] have studied the effects of HIPPV with peak airway pressure of 30 or 45 cm H_2O in rats. After one hour of 30 cm H_2O HIPPV the lung weight was increased attesting to the presence of edema. This was confirmed by histological examination that disclosed interstitial but not alveolar edema. As a matter of fact, the animals were not hypoxemic. By contrast, when submitted to a 45 cm H_2O HIPPV, the animals became very rapidly cyanotic, and were moribund or dead after less than 40 min. Indeed, profound hypoxemia was present together with tracheal flooding.

Pulmonary edema was very severe as demonstrated both by a nearly threefold increase in lung weight and the presence of marked alveolar edema.

What Is (Are) the Mechanism(s) of HIPPV Edema?

It had been postulated that this edema may be of the hydrostatic type, as the result of increased transmural pressure at the level of both alveolar and extra-alveolar

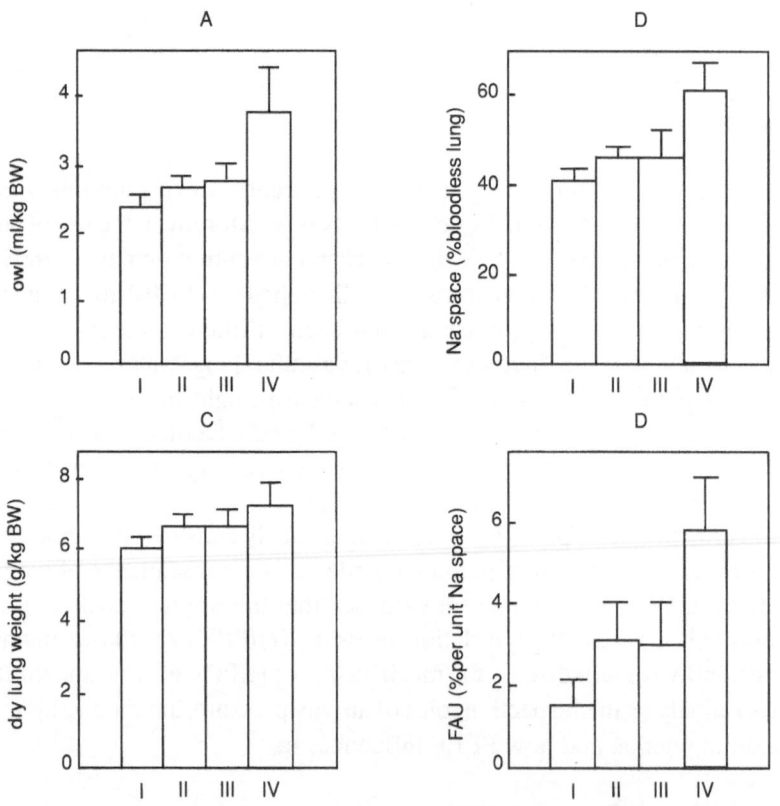

Fig. 1. Effects of HIPPV of increasing durations (I=Controls, II=5 min of HIPPV, III=10 min, IV=20 min) on the extravascular lung water (A), the distribution space of [22]Na in lungs (B), the dry lung weight (C), and the fractional albumin uptake (D). Significance of the differences from controls: *: p<.05, **: p<.01, ***: p<.001. Group IV differed from all others at least at p< .01. (Reproduced from Dreyfuss et al.[10] with permission)

microvessels.[7] Surfactant inactivation due to cyclic distention and compression of alveolar surface film[8] might have been responsible for the former, while the latter would have been the result of lung interdependence during inflation.[9]

Could epithelial/endothelial permeability changes be co-responsible for the occurrence of HIPPV edema? And if so, could cellular alterations be associated? In order to answer these questions, we performed experiments in rats ventilated with 45 cmH$_2$O HIPPV.[10]

The presence of permeability alterations was documented by increases in the dry lung weight and in the uptake of radio-labeled albumin by lungs (Fig. 1). These abnormalities were already significant after 5 or 10 min of HIPPV, and further increased after 20 min of HIPPV. At this stage, proteinaceous tracheal fluid was present in most of the animals. The variations of dry lung weight and albumin space paralleled the indexes of pulmonary edema (extravascular lung water and ^{22}Na distribution space in lungs).

A significant correlation was observed between extravascular lung water and dry lung weight with a slope that indicated the outpouring of pure plasma.[10] This high permeability edema was accompanied by cellular alterations culminating at 20 min of HIPPV. At this stage, diffuse alveolar damage was present.

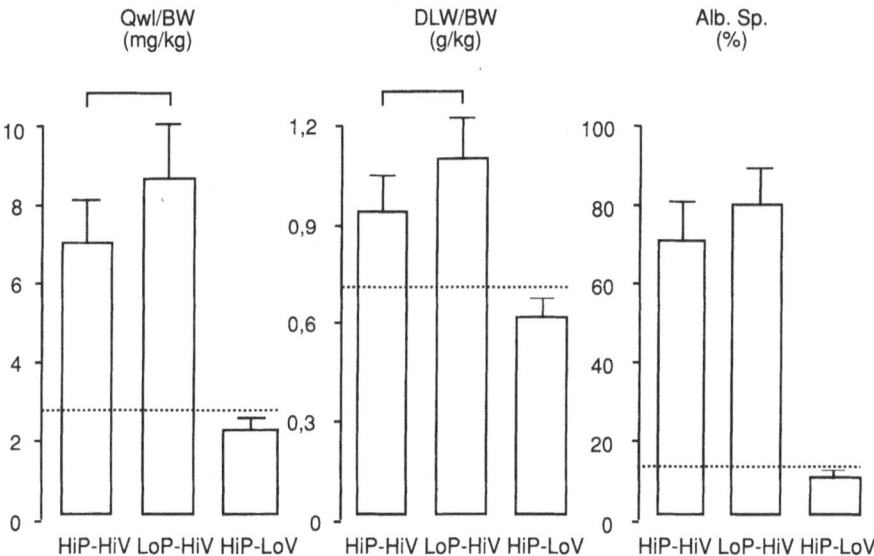

Fig. 2. Extravascular lung water, dry lung weight, and albumin space in rats ventilated with high airway pressure and high tidal volume (HiP-HiV), low airway pressure-high tidal volume (LoP- HiV) by means of an iron lung, and high airway pressure-low tidal volume (HiP-LoV) by means of thoraco-abdominal strapping. Horizontal dotted lines represent the upper 95% confidence limit for control values. HiP-HiV and LoP-HiV were always different from controls (p<. 001). **: p<. 01. (Reproduced from Dreyfuss et al.[11] with permission)

44

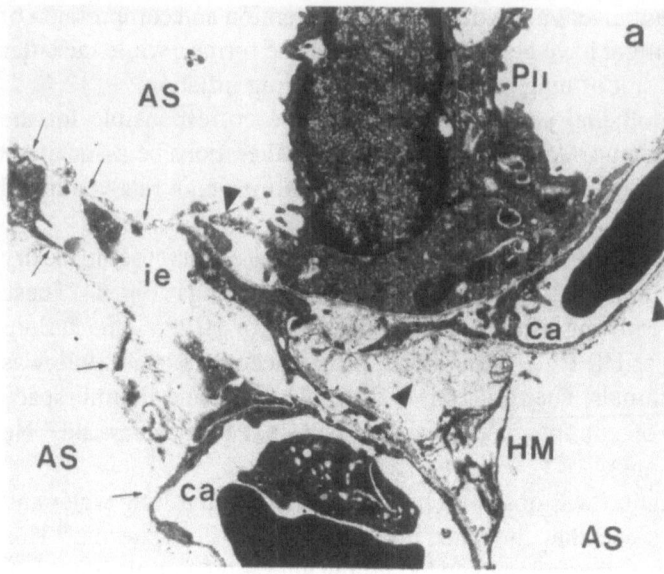

Fig. 3a. Ultrastructural lesions observed after high volume ventilation (whether airway pressure was high or low). There is lysis of type I cells (arrowheads) leading to denudation of basement membrane (arrows) and hyaline membrane formation (HM). Type II cells (PII) seem preserved. Interstitial edema (ie) is present. ca: capillary lumen, AS: alveolar space.

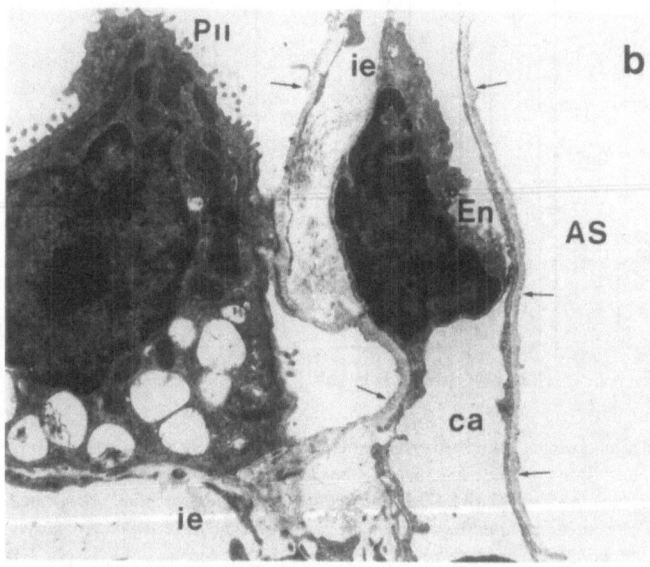

Fig. 3b. When PEEP is applied during HIPPV, there is preservation of the epithelial lining. En: endothelial cell. (Reproduced from Dreyfuss et al.[11] with permission)

What Are the Respective Roles of High Peak Airway Pressure and High Tidal Volume?

To address this point we made airway pressure and lung volume vary independently during mechanical ventilation in rats.[11] The effects of high peak airway pressure were dissociated form those of high lung volume by the utilization of three types of ventilatory modes.

High pressure-high volume ventilation was achieved with HIPPV 45 cm H_2O. High pressure-low volume ventilation was achieved with HIPPV 45 cm H_2O and reduced tidal volume by thoraco-abdominal strapping. An iron lung allowed low pressure-high volume ventilation. Permeability edema occurred in those animals submitted to lung hyperinflation, irrespective of the level of airway pressure (Fig. 2). By contrast, increasing airway pressure without lung hyperinflation did not result in edema and produced no tissue damage.

Does PEEP Interfere with Hyperinflation Edema?

The addition of a 10 cm H_2O PEEP to HIPPV 45 cm H_2O results in the same overall lung distention but in a smaller tidal volume. When rats were submitted to such a challenge, both edema[7-11] and permeability indexes[11] were diminished. Moreover, ultrastructural damage was less severe, with conservation of the integrity of the epithelial lining (Fig. 3).

This effect of PEEP on the amount of edema is surprising considering what is usually seen in other types of experimental permeability edema[2-3]. Moreover, a "protective effect" of PEEP on the epithelial lining was never observed. An unresolved issue is whether the epithelium was preserved because alveolar flooding did not occur (owing to reduction of the amount of edema), or whether preservation by PEEP of the epithelial layer impeded flooding since the epithelium constitutes the most important determinant of the low protein permeability of the alveolar-capillary barrier.[12]

In conclusion, acute pulmonary hyperinflation seems to be very deleterious to lungs. The possibility that lung structure and function alterations may worsen during mechanical ventilation of patients with acute respiratory failure should be considered when searching for optimal management of such patients.

References

1. Falke K.J., Pontoppidan H., Kumar A., Leith D.E., Geffin B., Laver M.B.: Ventilation with end-expiratory pressure in acute lung disease. J. Clin. Invest., 1972; 51: 2315-2323.
2. Demling R.H., Staub N.C., Edmunds H.L. Jr.: Effect of end-expiratory airway pressure on accumulation of extravascular lung water. J. Appl. Physiol., 1975; 39:672-9.

3. Toung T., Saharia P., Permutt S., Zuidema G.D., Cameron J.L.: Aspiration pneumonia: beneficial and harmful effects of positive end-expiratory pressure. Surgery 1977; 82: 279-283.

4. Iliff L.D.: Extra-alveolar vessels and edema development in excised dog lungs. Cir. Res., 1971; 28: 524-532.

5. Albert R.K., Lakshminaryan S., Kirk W., Butler J.: Lung inflation can cause pulmonary edema in zone I of *in situ* dog lungs. J. Appl. Physiol., 1980; 49: 815-819.

6. Rizk N.W., Murray J.F.: Peep and pulmonary edema. Am. J. Med. 1982; 72: 381-383.

7. Webb H.H., Tierney D.F.: Experimental pulmonary edema due to intermittent positive pressure ventilation with high inflation pressures. Protection by positive end-expiratory pressure. Am. Rev. Respir. Dis. 1974; 110: 556-565.

8. Faridy E.E., Permutt S., Riley R.L.: Effect of ventilation on surface forces in excised dogs' lungs. J. Appl. Physiol. 1966; 21: 1453-1462.

9. Permutt S.: Mechanical influences on water accumulation in the lungs. In: Fishman A.P., Renkin E.M. (Eds.), *Pulmonary Edema. Clinical Physiology Series*. The American Physiological Society, Bethesda. 1979; 175-193.

10. Dreyfuss D., Basset G., Soler P., Saumon G.: Intermittent positive-pressure hyperventilation with high inflation pressure produces pulmonary microvascular injury in rats. Am. Rev. Respir. Dis. 1985; 132: 880-884.

11. Dreyfuss D., Soler P., Basset G., Saumon G.: High inflation pressure pulmonary edema. Respective effects of high airway pressure, high tidal volume, and positive end-expiratory pressure. Am. Rev. Respir. Dis. 1988; 137: 1159-1164.

12. Staub N.C.: Pulmonary edema. Physiol. Rev., 1974; 54:678-811.

7. Is Pulmonary Hyperinflation a Feature in Advancing Interstitial Lung Disease?

R. TORCHIO, C. GULOTTA, G. SERA, D. BOARO, A. TOSADORI, C. BANAUDI
Respiratory Physiopathology Service, S. Luigi Hospital, Orbassano, Turin, Italy

Introduction

The end stage lung is the common feature of the diseases affecting the pulmonary interstitium and developing from mild degree to the very important architectural damage of the honey-combing lung.[1-4]

Arterial hypoxaemia develops in these conditions as a result of various mechanisms:[5-7] the reduction in the capillary bed, the thickening of the alveolar-capillary membrane, the ventilation perfusion mismatching, the low mixed venous oxygen tension due to low cardiac output and finally the shortened time available for diffusion along the capillary. The respiratory function studies show a ventilatory restrictive pattern in these conditions, but with advancing architectural damage, obstruction of small and peripheral airways can occur.[8-10] Pathological studies clearly show that in honey-combing lungs cystic areas and bullae are present in the lung parenchyma.[1-2]

Considering the pathological alteration present in interstitial lung disease, the aim of the present study was to establish if hyperinflation occurs during advancing interstitial lung disease and the effects on overall gas exchange using simple tests of respiratory function.

Patients and Methods

We studied ten healthy subjects (control group) and thirty patients: fifteen with a histological diagnosis of sarcoidosis (eight at chest X-ray stage II and seven at X-ray stage III; sarcoidosis group); fifteen with an advanced interstitial lung disease of various origin (I.L.D. group: two with progressive systemic sclerosis, three with asbestosis, ten with idiopathic pulmonary fibrosis). The basal anthropometric and functional data are given in Table I.

Table I. Basal anthropometric and functional data in the three groups ($\chi^2 \pm$ SD)

	Controls	Sarcoidosis	I. L. D.
n	10	15	15
males	5	6	8
females	5	9	7
age [years]	34.9±12.5	45.5±10.3	54.6±9.50
VC [% predicted]	87.1±9.06	83.4±19.7	64.8±20.6
FEV_1/VC [%]	91.7±10.9	77.7±11.8	81.4±19.5
FEF75 [% predicted]	100.±35.2	66.4±37.1	44.1±31.9
sGaw [kPa^-1.s^-1]	2.82±0.83	2.28±0.66	2.20±0.49
PaO_2 [mmHg]	102.±8.36	89.1±10.8	69.3±17.3

VC = Vital capacitys FEV_1 = forced expiratory volume at 1 sec
FEF75 = forced expiratory flow at 75% of forced vital capacity
sGaw = specific airway conductance

In the morning, between 9.00 and 12.00, the subjects performed a whole body plethysmography, a flow volume curve, a nitrogen wash-out for functional residual capacity (FRC) determination and a study of alveolar-arterial difference for oxygen ($AaPO_2$) and arterial-alveolar difference for CO_2 ($aAPCO_2$) with physiological (Vd/Vt) and alveolar (Vd/Vt alv) dead space measurement. European Coal and Steel Community predicted values were considered.

A Gould Bodybox plethysmograph was used to determine the thoracic gas volume (TGV), measured with the subject breathing at the tidal volume and putting the hands on the cheeks.[11] The nitrogen washout was performed with an open circuit registering the log of nitrogen concentration vs time. The FRC was calculated according to Darling and coworkers.[11]

The difference between TGV and FRC was calculated and expressed as absolute and as percentage of the TGV value. Alveolar-arterial differences and Vd/Vt measurements were performed with an original computer integrated system connecting an Airspec 200 MGA mass spectrometer and a spirometer to an HP 9816s computer. Samples of gases (nitrogen, oxygen, carbon dioxide) with the signal of changing volume were collected at a frequency of 1 every 50 milliseconds.

The delay between signals was taken into account and the calibration of signals was performed with linear regression on three points of different gas concentration.

The traces were graphically represented, on line, on the screen and the data were stored on a floppy disk.

Vd/Vt was calculated with the Enghoff correction[13] and Vd/Vt alv as:

$$[(PetCO_2 - PaCO_2) / PaCO_2]\ 100$$

Expired concentrations were obtained from areas under the curves, after the necessary corrections. The two components[13-14] of the alveolar-arterial difference for O_2 (difference between arterial and ideal alveolar PO_2; difference between ideal alveolar and end tidal PO_2) were calculated.

Results

The results are reported in Tables II and III and graphically represented in Figs. 1 and 2.

Table II. Parameters as indices of hyperinflation ($\chi^2 \pm$ SD)

	Controls	Sarcoidosis	I. L. D.
RV [% predicted]	95.4 ± 27.4	91.0 ± 20.4	92.3 ± 22.5
RV/TLC [%]	28.8 ± 4.80	31.7 ± 7.80	37.9 ± 6.50
RV/TLC [% predicted]	$109. \pm 23.7$	$113. \pm 37.8$	$126. \pm 22.6$
D TGV-FRC [I]	$.307 \pm .181$	$.383 \pm .265$	$.631 \pm .452$
D TGV-FRC [% TGV]	10.1 ± 6.30	13.0 ± 9.16	21.7 ± 15.2

RV = residual volume; TLC = total lung capacity; TGV = thoracic gas volume
FRC = functional residual capacity; D = difference

Table III. Measured parameters for gas exchange ($\chi^2 \pm$ SD)

	Controls	Sarcoidosis	I. L. D.
AaPO$_2$ [mmHg]	10.4 ± 4.40	24.1 ± 7.70	43.3 ± 23.0
PetO$_2$-PaO$_2$id [mmHg]	0.26 ± 0.60	1.22 ± 1.42	4.82 ± 2.71
aАPCO$_2$ [mmHg]	0.28 ± 0.90	1.01 ± 1.00	4.34 ± 2.80
Vd/Vt [%]	31.5 ± 5.01	31.2 ± 8.80	40.9 ± 11.9
Alveolar Vd/Vt [%]	0.88 ± 2.90	2.91 ± 2.90	12.2 ± 5.20

AaPO$_2$ = alveolar-arterial difference for O_2 PetO$_2$ = end tidal PO_2
aAPCO$_2$ = arterial-alveolar difference for CO_2 Vd/Vt = dead space
PAO$_2$id = ideal alveolar PO_2

The difference between TGV and FRC (D TGV-FRC), as absolute value, but also as percentage of TGV, was significantly higher in the I.L.D. group than in controls (t = 2.284 p<0.02) and in the sarcoidosis group (t = 1.909 p<0.05). The ratio between residual volume (RV) and total lung capacity (TLC), shows the same pattern (t = 3.760 p<0.001 I.L.D. group vs controls and t = 2.350 p<0.02 I.L.D. versus sarcoidosis). These findings remain the same whether considering the RV,' TLC ratio as percentage of the predicted or as absolute values). No statistically significant differences were observed between the control group and the sarcoidosis group for these parameters. Significantly lower values of forced expiratory flow at 75% of forced vital capacity (FEF75) were observed between the control and the sarcoidosis groups (t = 2.309 p<0.02). The alveolar-arterial difference for O_2 (AaPO$_2$) was significantly higher than in controls in sarcoidosis (t = 5.020 p<0.001) and in the I.L.D. group (t= 4.270 p<0.001), but the arterial-alveolar difference for

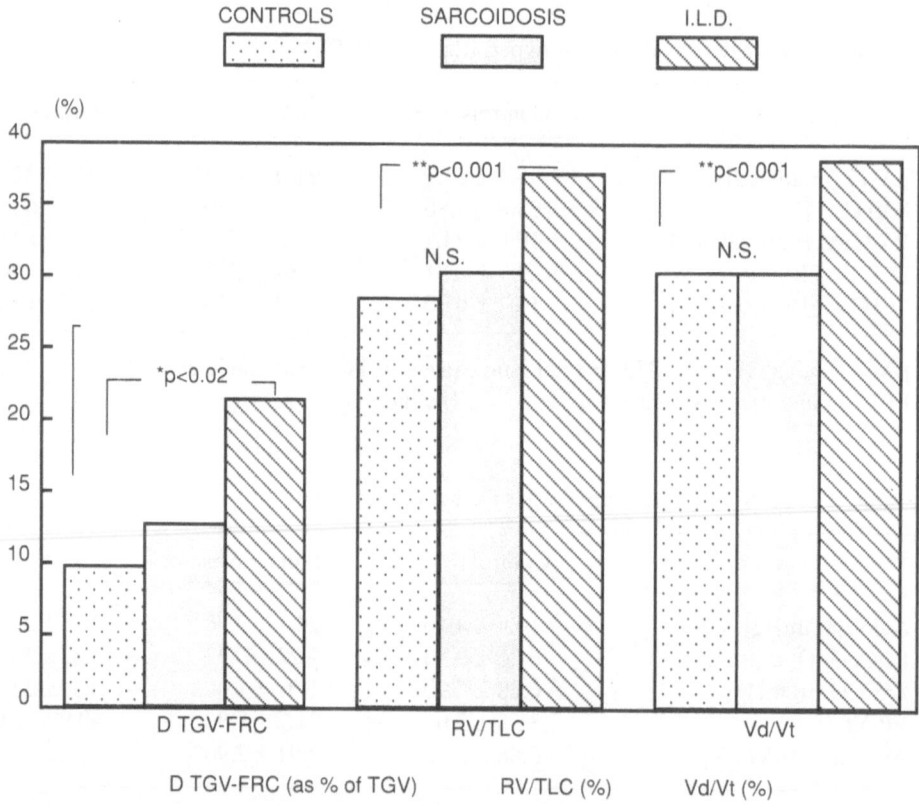

Fig. 1. Parameters of hyperinflation in the three groups [difference between thoracic gas volume and functional residual capacity (D TGV-FRC), residual volume - total lung capacity ratio (RV/TLC), and physiological dead space] with statistical analysis.

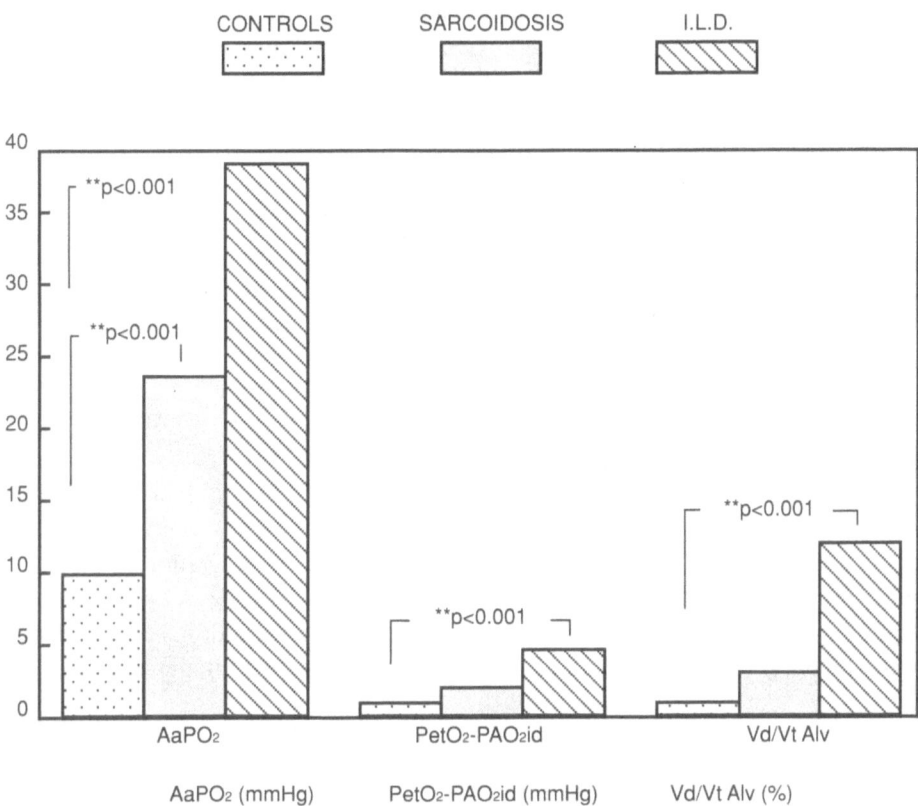

Fig. 2. Alveolar-arterial difference for O_2 (AaPO$_2$), difference between end tidal and alveolar ideal PO$_2$ (PetO$_2$-PaO$_2$id) and alveolar dead space in the three groups with statistically significant differences.

CO_2 (aAPCO$_2$) was in the normal range in the sarcoidosis but significantly higher in the I.L.D. group (t = 4.270 p<0.001 versus controls). Finally a slight but statistically significant correlation was present between D TGV-FRC and aAPCO$_2$ (r = .360 p<0.05), D TGV-FRC and Vd/Vt alv (r = .375 p<0.05); D TGV-FRC and the difference between end tidal PO$_2$ and ideal alveolar PO$_2$.

Discussion

In the subjects studied a normal function in central airways was observed (sGaw, Raw). The ratio of FEV$_1$ to VC (FEV$_1$/VC) was in the normal range, though lower than in controls, in the I.L.D. group, showing a substantial integrity in airways proximal to the equal pressure point. An alteration was present, however, in terminal forced expiratory flows in all the patients with interstitial lung disease, clearly showing mechanical damage in the small airways. All the patients in the

I.L.D. group showed alterations in the parameters of hyperinflation. Furthermore, the presence of a high D TGV-FRC can be considered[15-16] indicative of the presence of poorly accessible compartments of alveoli. The D TGV-FRC higher than in controls in the I.L.D. group, but not in stage II and III sarcoidosis, may be related to the presence of bullae or non communicating cysts[2] present in patients with honey-combing lung. According to the literature[5-6] in these patients the hypoxaemia is principally related to the presence of alveoli with low $\dot{V}/\dot{Q}$ ratio, but at rest[5-7] the presence of high $\dot{V}/\dot{Q}$ alveoli was also described associated with advanced stages of disease. Our findings - high $aAPO_2$, high difference between end tidal PO_2 and ideal alveolar PO_2 (presence of high $\dot{V}/\dot{Q}$ alveoli), high arterial alveolar dead space - agree with these previous observations.

Conclusions

In advancing interstitial lung diseases an obstruction in small airways appears and some degree of pulmonary air trapping, with normal absolute valve of RV, occurs. The alveolar-arterial difference for O_2 becomes more and more pronounced with progression of the disease, and in the more advanced stages of the disease a significant dead space ventilation may be present. These findings are consistent with the progressive architectural distortion present in the lung parenchyma.

References

1. Petty T.L., Filley G.F. (Eds.):Interstitial lung disease (18th Aspen Lung Conference): Chest 1976; 69: 2S, 251-325.
2. Crystal R.G., Fulmer J.D., Roberts W.C., Morton L., Moss M.D., Line B.R., Reynolds H.Y.: Idiopathic pulmonary fibrosis. Clinical histological, radiographic, physiologic, scintigraphic, cytologic and biochemical aspects. Ann. Intern. Med. 1976; 85: 769-788.
3. Schatz M., Patterson R., Fink J.: Immunologic lung disease. New Engl.J.Med. 1979; 300, 23: 1310-1320.
4. Gupta S.K.: Sarcoidosis: clinical aspects. State of the art. In: Grassi C. Rizzato G. (Eds.) *Sarcoidosis and other granulomatous disorders* . Amsterdam, Elsevier, 1988, 397-406.
5. Jernudd-Wilhelmsson Y., Hornblad Y. Hedenstierna G.: Ventilation-perfusion relationships in interstitial lung disease. Eur. J. Respir. Dis 1986; 68: 39-49.
6. Wagner P.D., Dantzker D.R., Dereck R., De Polo J.L., Wasserman K., West J.B.: Distribution of ventilation perfusion ratios in patients with interstitial lung disease. Chest 1976; 69: 256-257.
7 Prediletto R., Formichi B., Viegi G., Fornai E., Begliomini E., Santolicandro A., Giuntini C.: Multiple inert gas elimination technique in interstitial lung disease: analysis of ventilation-perfusion relationships. In: Grassi C., Rizzato G. Pozzi E. (Eds.): *Sarcoidosis and other granulomatous disorders.* Amsterdam, Elsevier, 1988, 375-376.
8. Benson M.K., Hughes D.T.: Serial pulmonary function tests in fibrosing alveolitis. Br. J. Dis. Chest 1972; 66: 33-44

9. Miller R A., Teirstein A.S., Pilkipski M., Brown L.K.: The spectrum of airways obstruction in sarcoidosis. In: Grassi C., Rizzato G., Pozzi E. (Eds.) *Sarcoidosis and other granulomatous disorders,* Amsterdam, Elsevier, 1988, 351-354.

10. Ostrow D., Cherniack R.M.: Resistance to airflow in patients with diffuse interstitial lung disease. Am. Rev. Resp. Dis. 1973; 108: 205-210.

11. Darling R.C., Cournand A., Richard D.W. Jr: Studies on the intrapulmonary mixture of gases. An open circuit method for measuring residual air. J. Clin. Invest. 1940; 12: 609-616.

12. Du Bois A.B., Bothelho S.Y., Bedel G.N., Maishall R., Comroe J.E. Jr.: A rapid plethysmographic method for measuring thoracic gas volume comparison with nitrogen washout method for measuring functional residual capacity in normal subjects. J. Clin. Invest. 1956: 34: 322-326.

13. Anthonisen N.R., Fleetham S.A.: Ventilation: total, alveolar and dead space. In: Fahri L.E., Tenney S.M. (Eds.): *Handbook of physiology*, Bethesda, Am. Physiol. Soc. 1987; 4: 113-129.

14. Fahri L.E.: Ventilation perfusion relationships. In: Fahri L.E., Tenney S.M. (Eds.), *Handbook of physiology*, Bethesda, Am. Physiol. Soc. 1987; 4: 199-215.

15. Matthys H. In: *Lungenfunktiondiagnostik mittel Ganzkorperplethysmographie*. Stuttgart Schatthauer Verlag, 1972

16. Herzog H., Keller R., Amrein R., Matthys H., Joos J.: Patterns of correlation of pulmonary function values determined by spirography and body plethysmography. Prog. Resp. Dis. 1969; 14: 205-214.

Physiopathological Effects

8. Effects of Lung Hyperinflation on Pulmonary Circulation

C. EMERY
Department of Medicine, Royal Hallamshire Hospital, Sheffield, UK

Introduction

Inflation of the lung affects large and small pulmonary vessels in two distinct ways. It is thought to dilate extraalveolar vessels (large arteries and veins) as perivascular pressure becomes more negative but to compress and stretch alveolar vessels (located in the alveolar wall and surrounded by alveolar pressure). Under West's Zone II conditions, where pulmonary artery pressure (Ppa) exceeds alveolar pressure (Palv) which in turn is greater than left atrial pressure (Pla), blood flow (Q) depends on the pressure gradient Ppa-Palv with Palv being the effective downstream pressure. When Palv exceeds Pla, thin walled vessels subjected to alveolar pressure may act as Starling resistors; that is they flutter open and shut with flow like the device used by Starling in his heart-lung preparation. Permutt and Riley [1] showed that small muscular vessels could also act as Starling resistors as they collapse due to muscular tone or critical closing pressure. In chronic hypoxia the development of new muscle in normal thin walled arterioles down to 20 μm diameter and surrounded by alveoli suggests that these vessels may collapse due to muscle tone. There is also narrowing of the lumen of these vessels due to the encroachment by the new muscle. Could this affect the influence of alveolar pressure on the pulmonary circulation? Harris et al.[2,3] observed that during exercise Ppa/Q lines were displaced upwards to higher Ppa values in chronic bronchitic patients but not in normal subjects. They suggested that this might be due to raised alveolar pressure in bronchitic patients. We suggest that the newly muscularised arterioles in chronic hypoxic lung may alter the effect of raised Palv on the pulmonary circulation. We compared the effect of hyperinflation of the lung on the pulmonary circulation of control and chronically hypoxic rats during normoxia and during hypoxia. The effect of a potent vasodilator, NECA, given during hypoxic vasoconstriction was also studied.

Materials and Methods

The effect of alveolar pressure on the pulmonary circulation was studied in both chronically hypoxic (CH) and littermate control (C) male Wistar rats. Litters of 3-4 week old rats were split and half were placed in a normobaric hypoxic chamber maintained at 10% O_2, for 2-3 weeks. Isolated lungs were perfused in situ with a roller pump and ventilation maintained with air + 5% CO_2.

The lungs were perfused with homologous blood at 38°C, pH 7.35-7.45, at a constant flow rate of 20 ml/min^{-1} through the cannulated pulmonary artery. Blood escaped from the cannulated left atrium to a heated reservoir and was then recirculated. Acute hypoxia was achieved by ventilation of the lung with 2% O_2 + 5% CO_2. Drugs were injected into the main pulmonary artery through rubber tubing. Pulmonary artery pressure, left atrial pressure, airway pressure (=alveolar pressure) and flow rate (electromagnetic flowmeter) were measured. End expiratory pressure was 2-3 cmH$_2$O.

Pressure/ Flow Curves (Ppa/Q)

Blood flow was slowly reduced from 25 or 20 ml/min to zero by altering the output of the perfusion pump. The relationship between Ppa and flow was recorded on an X-Y recorder with flow on the abscissa and pressure on the ordinate. The left atrial cannula was positioned so that left atrial pressure was zero or sub zero. Palv was thus higher than Pla and the lungs were under Zone II conditions. Before plotting Ppa/Q relations, rhythmic ventilation was interrupted and the lungs inflated to a set Palv (usually 5 or 15 mmHg).

Ventilating gas mixture was blown across a T piece whose vertical limb was immersed in water to a depth equivalent to the required Palv; excess gas escaped from this trap. Thus Ppa/Q lines could be plotted during both normoxic and hypoxic inflation. Ppa/Q lines were measured when the lungs were in a steady state. They were linear over a wide range but curved convexly to the pressure axis at very low flow rates. The linear part was extrapolated to the pressure axis to form an intercept. The slope of the Ppa/Q line represented vascular resistance, while the intercept pressure was attributable to vessels acting as "Starling resistors" and forming the effective downstream pressure for flow. Flow over the linear portion was proportional to Ppa-Intercept pressure.

Ppa/Q Lines During Vasoconstriction by Hypoxia (HPV)

Ppa/Q lines were plotted during normoxia. Ventilation with a low O_2 mixture (2%

$O_2 + 5\%$ CO_2) caused a rise in Ppa. Ppa/Q lines were plotted when this raised Ppa was stable.

P/Q Lines during Reversal of HPV by 5 μg n-Ethyl Carbamino Adenosine (NECA)
    Ppa/Q lines were plotted during hypoxia as above. After administration of vasodilator Ppa fell to near normoxic levels. When Ppa was stable Ppa/Q lines were plotted.

Results

Lung Inflation
    In C rats high Palv (15 mmHg) caused a rise in Ppa which then declined to a steady level; after deflation Ppa fell rapidly below the initial value and then rose (Fig. 1). In CH rats high Palv caused an initial rise in Ppa and then a further climb; after deflation Ppa fell rapidly first and then more slowly towards the control level, falling below this level.

Ppa/Q Curve Normoxia
    In both C and CH rats raising Palv from 5 to 15 mmHg caused a parallel upward displacement of the Ppa/Q line on the pressure axis approximating to the change in Palv. The slope and extrapolated intercept of the line were significantly greater in CH lungs than in C lungs. Traces of typical lines are shown in Fig. 2.

Hypoxia
    During hypoxia the Ppa/Q lines were shifted up the pressure axis with an increase in slope and intercept. In C rats raising Palv from 5 to 15 mmHg caused an increase in intercept which was less than the change in Palv whilst in CH rats the increase in intercept was greater than the change in Palv, as seen in Fig. 2. Fig. 3 shows mean Ppa/Q lines measured during normoxia and hypoxia in 6 C and 6 CH rats; they show the same characteristics as the individual lines in Fig. 2.

Reversal of HPV with NECA
    After plotting Ppa/Q lines during hypoxic vasoconstriction 5 μg NECA was administered into the circuit. This dose almost completely abolished the vasoconstriction caused by hypoxia as shown in Fig. 3. In both C and CH rats the intercept and slope returned close to normoxic values during continued hypoxia.

Discussion

    There is evidence in both animal and man that a raised airway or inflation pressure causes a rise in Ppa and a parallel shift in the pressure/flow line to higher

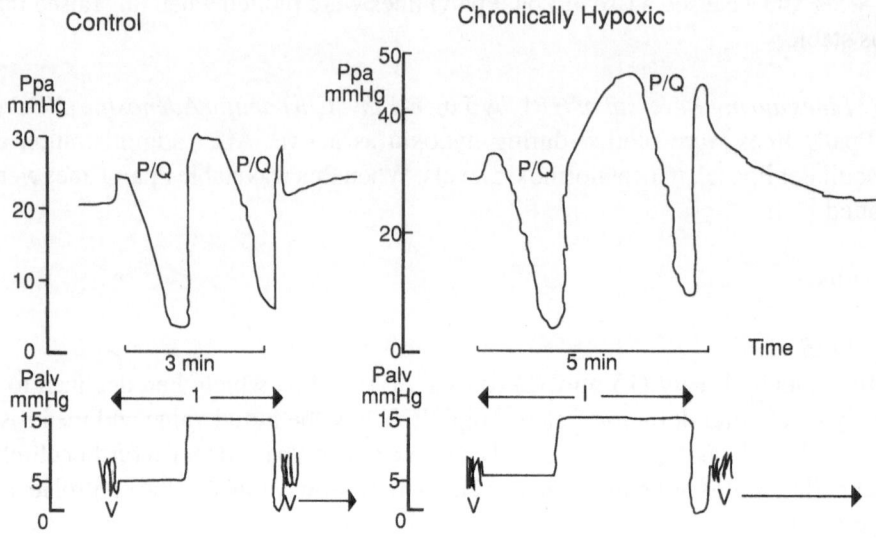

Fig. 1. Effect of inflation on Ppa in control and chronically hypoxic rat (trace). Upper trace is Ppa and lower trace is corresponding Palv. V = rhythmic ventilation, I = inflation of lung to set Palv, low (5 mmHg) or high (15 mmHg) pressure; ventilation stopped. P/Q = reduction of flow at constant Palv to draw Ppa/Q lines on X-Y recorder. Flow is then restored to 20ml/min^{-1}. Inflation of the lung raises Ppa by approximately *Palv in C lung but >*Palv in CH lung at high inflation pressure.

pressures for similar flow rates. De Bono and Caro[4] demontrated this shift in dog lungs and Sylvester and colleagues[5] made similar observations in pig lungs. Figures 2 and 3 show the same phenomenon during normoxia in our rat lungs. Harris and colleagues [2,3] were able to measure two Ppa/Q points in man by assuming 50% of cardiac output went through one lung in the control state and then doubling that flow by occlusion of one pulmonary artery. In normal man the two points were displaced in a "parallel" fashion when they breathed against an expiratory resistance. They then measured Ppa/Q points in normal subjects and in two groups with pulmonary hypertension, chronic bronchitic and mitral stenotic patients, at rest and during exercise. The "exercise points" lay on a continuation of the line measured at rest in normals and mitral stenotic patients. However, during exercise in chronic bronchitic patients the points were displaced up the pressure axis, similar to the effect of raised airway pressure in normal man.

We wondered whether the muscularised arterioles of hypoxic chronic bronchitic patients might play a part in this phenomenon. Our aim was to see whether chronic hypoxia altered the influence of alveolar pressure by studying the effect of hyperinflation in control and chronically hypoxic rat lungs. Young rats develop cardio-pulmonary changes which resemble those found in chronically hypoxic

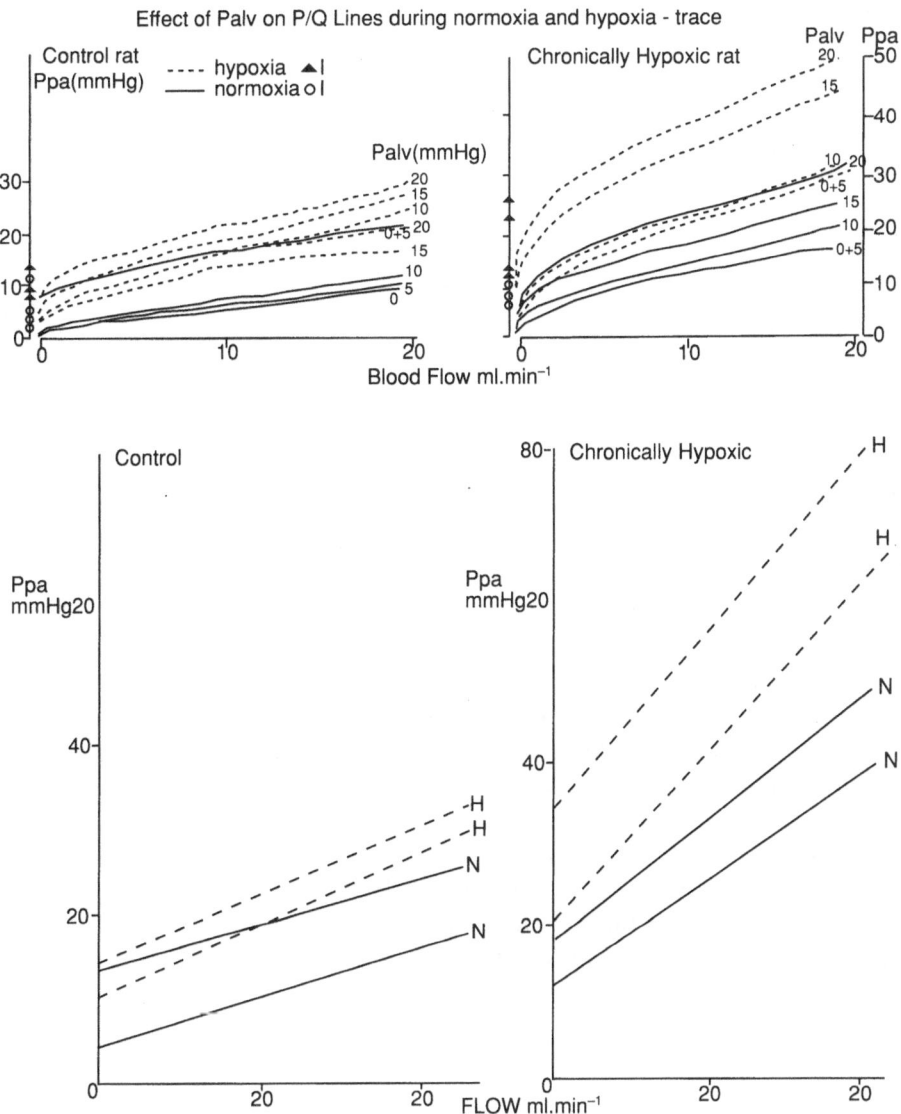

Fig. 2. Calculated regression lines of Ppa/Q lines from a typical control and chronically hypoxic rat isolated lung during normoxia (N—) and hypoxia (H......) at 5 and 15 mmHg (lower and upper lines respectively).

man, whether due to high altitude or chronic hypoxic lung disease.[6]

By measuring Ppa/Q relations in these isolated lungs at low and high Palv we were able to evaluate changes in both resistance and "Starling resistor" properties of the pulmonary circulation.

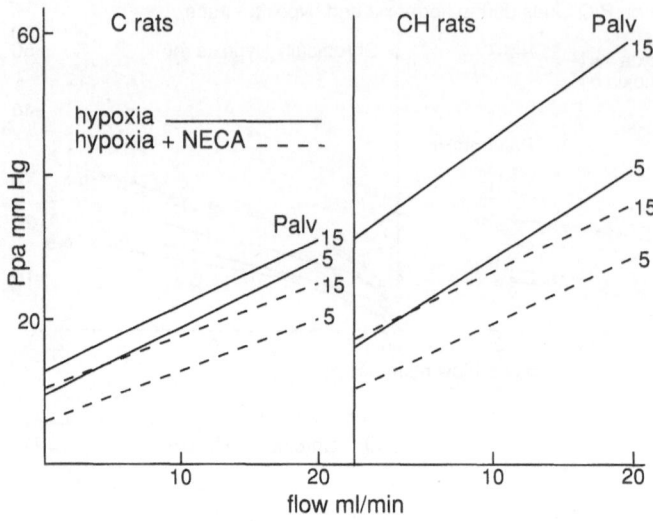

Fig. 3. Calculated mean regression lines of Ppa/Q lines in 6 C and 6 CH rats during hypoxia (——) and after 5 μg NECA during hypoxia (......) at 5 and 15 mmHg Palv. Normoxic lines were as in Fig. 2. NECA restored Ppa/Q lines to near normoxic values.

The Effect of Hyperinflation on Pulmonary Artery Pressure

Inflating the lungs from 5 to 15 mmHg at constant blood flow caused an increase in Ppa which in C rats declined thereafter to a lower level. In CH lungs Ppa rose sharply and then continued to climb at a slower rate to a level much greater than the rise in Palv. Thus in C rats there was some dilatation after rapid inflation whilst in CH rats inflation seemed to provoke vasoconstriction. On deflation similar but opposite changes occurred. Measurement of Ppa/Q lines showed that both Starling resistor and resistance values are raised in CH rats which we attribute to normoxic tone in the newly muscularised alveolar vessels. Raising Palv from 5 to 15 mmHg caused a parallel shift in the lines in both groups. The increase in intercept was near equivalent to the change in alveolar pressure which suggested full transmission of Palv to the pulmonary artery.

The Effect of Hyperinflation on the Pulmonary Vasculature Constricted by Hypoxia

Hypoxic ventilation caused a rise in pulmonary artery pressure in both groups. The Ppa/Q lines were raised in both C and CH lungs with an increase in slope, indicating a raised pulmonary vascular resistance, and an increase in intercept, indicating the development of a new Starling resistor due to muscle tone. In C lungs raising Palv from 5 to 15 mmHg during hypoxia caused a small shift in the Ppa/Q line with the rise in intercept being much less than the rise in Palv (Figs. 2,3). Thus the transmission of Palv was greatly reduced. However in CH lungs the rise in intercept was greater than the change in Palv. Thus Palv was fully transmitted to Ppa and in some way enhanced vascular tone. We have earlier suggested that this difference in the influence of inflation on the pulmonary circulation between C and CH lungs is due to the presence of newly muscularised arterioles in the alveolar region of CH lungs.[7] We provided evidence for a peripheral shift in the major site of hypoxic vasoconstriction to these vessels in chronic hypoxia. In C rats the major site of vasoconstriction is in small muscular vessels upstream from alveolar vessels, thus forming a Starling resistor in the extra-alveolar region which becomes the effective downstream pressure. Alveolar pressure has to overcome this new resistor before it can influence pulmonary artery pressure. In unpublished work it was found that Palv had to be raised to 15-20 mmHg before Ppa/Q lines were shifted. In CH rats the major site of vasoconstriction is in the newly muscularised alveolar vessels which form a new Starling resistor. The raised tone in these vessels summates with, and is enhanced by, the rise in alveolar pressure as indicated by the shift of the Ppa/Q line being greater than the rise in Palv. It probably originates in a geometric change in the vessels.

We have shown that vasoconstriction with almitrine affects the influence of lung inflation on the pulmonary circulation in a similar manner to hypoxia.[8] Almitrine caused a rise in Ppa; raising Palv from 5 to 15 mmHg caused a small upward displacement of the Ppa/Q line in C rats but an exaggerated shift in CH rats. Thus almitrine appears to constrict the same vessels as hypoxia in both control and chronically hypoxic lungs.

Ppa/Q Lines After Reversal of Hypoxic Vasoconstriction by NECA

After the administration of the potent pulmonary vasodilator NECA (when the rise in Ppa with hypoxia was stable) Ppa fell to near normoxic levels. Ppa/Q lines were returned to near normoxic levels; both slope and intercept were reduced. In C and CH lungs full transmission of Palv was observed on raising Palv from 5 to 15 mmHg even though the lungs were still hypoxic. Thus it appears that NECA reduced tone in both extra- and alveolar resistors as well as resistance in larger vessels. Unpublished work by Professor Cai in this department has shown that another new potent vasodilator, Ligustrazine (derived from a Chinese herbal remedy) had effects similar to NECA. However, neither drug returned Ppa/Q lines in CH rats to the values found in C rats. The remaining differences after dilatation are attributable to structural changes in the pulmonary vessels of CH rats.

Implications for Chronically Hypoxic Patients

Alveolar pressure is normally low but may be raised in COPD patients. This is particularly likely during exercise, forced expiration or during periods of artificial ventilation. In normal subjects raising Palv raises Ppa probably due to increasing surrounding pressure of alveolar vessels. Chronically hypoxic rat lungs show similar results to control rat lungs although their initial resistor and resistance values are higher. However, if this is coupled with constriction of newly muscularised alveolar vessels the transmission of Palv is exaggerated as their increased tone summates and is augmented by the change in Palv; in control lungs the development of a higher resistor upstream from alveolar vessels "protects" the pulmonary artery from rises in Palv. We have evidence that the chronically hypoxic pulmonary circulation is more reactive to intrinsic vasoconstrictors and dilators [9,10] including angiotensins I and II, ATP and bradykinin as well as hypoxia. Thus the pulmonary circulation is likely to be in an increased state of tone in chronically hypoxic patients. Hyperinflation of the lung may thus have a marked effect on pulmonary artery pressure. Indeed, it may account for the dramatic rise in Ppa observed during exercise in some chronically hypoxic patients. The search for a specific pulmonary vasodilator for the treatment of pulmonary hypertension is important. The vasodilators we have tried reduced Ppa during hypoxic vasoconstriction in both C and CH rat lungs. This had the advantage of reducing the transmission of Palv to Ppa in hypoxic CH lungs although it increased transmission in C rats. Vasodilators so far tried do not reduce Ppa down to normal levels either in man or experimental animals. Only long term elevation of blood oxygen levels is successful in this respect. Hyperinflation of the lung affects the pulmonary circulation by transmission of raised alveolar pressure to pulmonary artery pressure under Zone II conditions. High tone in alveolar vessels of chronic hypoxia may push more vessels into Zone II conditions with consequences for alveolar capillary flow. More investigations in human chronically hypoxic patients are necessary.

References

1. Permutt S., Riley R.L.: Hemodynamics of collapsible vessels with tone: the vascular waterfall. J. App. Physiol. 1963; 18: 924-932.
2. Harris P., Segel N., Bishop J.M.: The relationship between pressure and flow in the pulmonary circulation in normal subjects and in patients with chronic bronchitis and mitral stenosis. Cardiovasc. Res. 1968; 2: 73-83.
3. Harris P., Segel N., Green I., Housley E.: The influence of airways resistance and alveolar pressure on the pulmonary vascular resistance in chronic bronchitics. Cadiovasc. Res. 1968; 2: 84-92.
4. De Bono E.F., Caro C.G.: Effect of lung inflating pressure on pulmonary blood pressure and flow. Am. J. Physiol. 1963; 205: 1178-86.
5. Sylvester J.T., Mitzner W., Ngeon Y., Permutt S.: Hypoxic constriction of alveolar and extra-alveolar vessels in isolated pig lungs. J. Appl. Physiol. 1983; 54: 1660-1666.
6. Leach E., Howard P., Barer G.R.: Resolution of hypoxic changes in the heart and pulmonary arterioles of rats during intermittent correction of hypoxia. Clin. Sci. Molec. Med. 1977; 52: 153-162.

7. Wach R.W., Emery C.J., Bee D., Barer G.R. Effect of alveolar pressure on pulmonary artery pressure in chronically hypoxic rats. Cardiovasc. Res. 1987; 21: 140-150.

8. Barer G.R., Cai Y.N.: The site of almitrine and hypoxic pulmonary vasoconstriction moves distally in chronically hypoxic rats associated with new growth of muscles in arterioles. J. Physiol. in press.

9. Emery C.J., Bee D., Barer G.R.: Mechanical properties and reactivity of vessels in the isolated perfused lungs of chronically hypoxic rats. Clin. Sci. 1981; 61: 569-580.

10. Russell P.C., Emery C.J., Cai Y.N., Barer G.R.: Enhanced reactivity of pulmonary vessels to Angiotensin I and Bradykinin in chronically hypoxic rats. Clin. Sci. in press.

9. Effects of Positive End-Expiratory Pressure (PEEP) on Bronchial Blood Flow

P. Agostoni, E. Doria, M. Pepi, G. Tamborini
Institute of Cardiology, Institute of Cardiovascular Research "G. Sisini", C.N.R., Cardiological Center "Fondazione I. Monzino", University of Milan, Italy

Introduction

The bronchial circulation is the source of systemic blood to the lung. The drainage of bronchial blood flow to structures outside the lung (trachea, main bronchi, esophagus, etc.) is into the right heart.[1-3] Conversely, the drainage of bronchial blood flow to the lung (lung parenchyma and intrapulmonary bronchi) is via the broncho-pulmonary anastomoses into the pulmonary circulation.[1-3] Therefore, this portion of bronchial blood flow is named systemic to pulmonary bronchial blood flow [$\dot{Q}br(s-p)$].[4]

$\dot{Q}br(s-p)$ has been measured in animal preparations[2,3,5,6] and found to be very sensitive to body temperature and body temperature changes.[3,5] Under controlled temperature conditions $\dot{Q}br(s-p)$ is ~ 60 ml/min/100 gdlw (g: grams, d: dry, l: lung, w: weight) in the anesthetized dog.[3,5]

Baile et al. were able to measure $\dot{Q}br(s-p)$ in humans during total cardio-pulmonary bypass.[7] However, because during total cardio-pulmonary bypass body temperature is unphysiologically low (range 26-31°C) these measurements of $\dot{Q}bs(s-p)$ should be considered with caution. $\dot{Q}br(s-p)$ is important for the metabolic needs of the lung parenchyma particularly in lung diseases. For instance, it has been reported that in lung injury $\dot{Q}br(s-p)$ has a fourfold increase[8] and it has been suggested that $\dot{Q}br(s-p)$ prevents pulmonary infarction after pulmonary embolism.[9] Indeed, $\dot{Q}br(s-p)$ is the major source of blood to the lung parenchyma after pulmonary embolism and it has been shown that when $\dot{Q}br(s-p)$ is reduced, as when pulmonary venous pressure is increased,[6] pulmonary infarction is more likely to occur as a consequence of pulmonary embolism.[9,10] Positive end-expiratory pressure (PEEP) is frequently utilized in patients with respiratory failure. It has been suggested in different animal species that $\dot{Q}br(s\text{-}p)$ is significantly reduced by PEEP.[2,11-13] This observation should imply a possible negative effect of PEEP on the survival of the lung parenchyma. However, temperature control of these experimental settings was, at least, uncertain. The present studies were undertaken to answer the following questions:

1) Is $\dot{Q}br(s\text{-}p)$ reduced by PEEP in the dog in a carefully temperature controlled preparation?

2) Is $\dot{Q}br(s\text{-}p)$ reduced by PEEP in humans during total cardio-pulmonary by-pass when body temperature is necessarily low?

Animal Studies

Methods

This study was performed in the laboratories of the University of Washington (Department of Respiratory Diseases). In 10 fully anesthetized dogs $\dot{Q}br(s\text{-}p)$ was measured using a previously described technique.[3,5,6]

In brief, the thorax of the dog was wide open, and the left upper lobe and the lingula were excised. The left lower lobe (LLL) and the right lung were ventilated separately with two ventilators and a double lumen tracheal tube. The LLL was hung in a fabric net on a strain gauge so that its weight changes could be recorded. The LLL pulmonary artery and vein were isolated.

The LLL circulation was sustained via an extracorporeal circuit which included, in sequence, a glass cannula from the pulmonary vein, a vascular reservoir with an overflow cannula open to the atmosphere at the level of the lowermost portion of the LLL, a roller pump, a coil of tubing and a glass cannula inserted in the pulmonary artery.

Both the LLL vascular reservoir and the coil of tubing were immersed in a waterbath the temperature of which was set to equalize LLL pulmonary artery temperature to the animal's core temperature. The back of the dog was thermally isolated from the operating table with an isolating pad. To avoid loss of heat from the open chest, the thorax and the abdomen of the dog were enclosed in a plexiglass box. The temperature inside the box was matched with the dog core temperature ($\pm$ 0.5° C) and the air was 100% humidified.

We recorded continuously right lung and LLL inspired gas temperature (30° C), right pulmonary artery temperature (used as dog body temperature), LLL pulmonary artery temperature and the temperature of the air inside the box.

$\dot{Q}$br(s-p) was measured as the volume of blood collected from the LLL vascular reservoir overflow cannula. To account for LLL vascular volume changes were added to the overflow volume.

Study Protocol and Statistical Analysis

After $\dot{Q}$br(s-p), hemodynamics, temperatures and ventilatory status of the dog had been stable for at least 30 minutes, $\dot{Q}$br(s-p) was measured for 15 minutes with PEEP = 5 cm H_2O and for a further 15 minutes with PEEP = 15 cm H_2O. The data for each 15 minute period were averaged. To compare the effects of PEEP on $\dot{Q}$br(s-p) in the dog with those in humans the data are reported as % changes in relation to control (PEEP = 5 cm H_2O).

Results

The increase in LLL PEEP from 5 to 15 cm H_2O reduces $\dot{Q}$br(s-p) by 49 % (Fig. 1). This $\dot{Q}$br(s-p) reduction was associated with an increase in LLL pulmonary artery pressure from 18 ± 2 cm H_2O to 32 ± 2 cm H_2O (mean $\pm$ SE; LLL pulmonary vein pressure was constantly ~ 0 cm H_2O because the overflow cannula was open to the atmosphere at the level of the lowermost portion of the LLL.

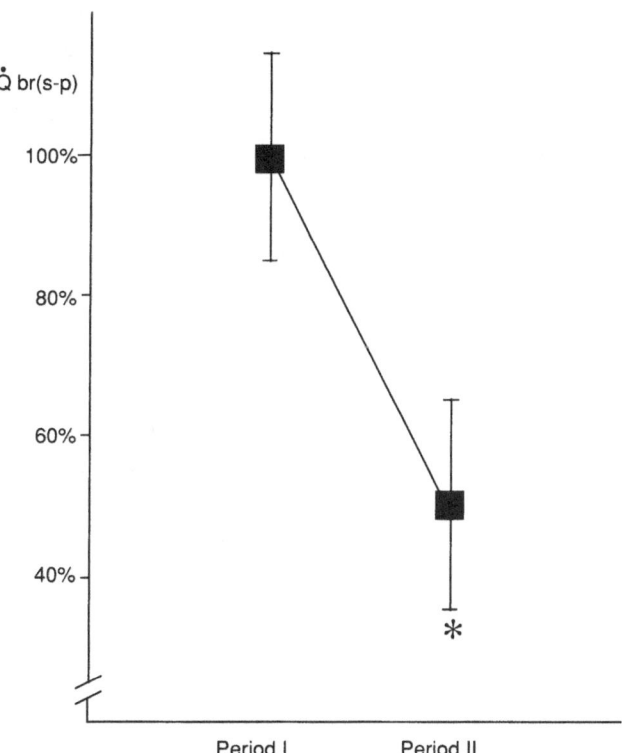

Fig. 1. Animal studies. Qbr(s-p) changes (% in relation to period I). Period I PEEP = 5 cm H_2O, Period II PEEP = 15 cm H_2O.

Human Studies

Methods

We studied 16 patients who had undergone total cardio-pulmonary bypass needed because of coronary artery bypass surgery. None had lung diseases or congenital malformations. The patients were 60.6 ± 5.4 years old (mean $\pm$ SD). All the patients were smokers. Total cardio-pulmonary bypass and systemic cooling of the patients were achieved using a standard technique. The pulmonary artery and the aorta were clamped. Patients were not ventilated. Systemic blood pressure, cardiac output (pump flow), rectal and esophageal temperatures were continuously recorded.

$\dot{Q}$br(s-p) was measured as the volume of blood returning to the left heart using a previously described technique.[7] A double lumen cannula was placed in the lower most portion of the left heart (located in the left atrium because the surgeon usually lifts the cardiac apex to reach the cardiac posterior wall) via the right superior pulmonary vein. One of the two lumens (8 French) was open to avoid the presence of negative pressure in the left atrium. The second lumen of the atrial cannula (18 French) was connected to a calibrated cylinder located ~ 50 cm below the left atrium at whose end there was a stopcock. From the cylinder the blood was propelled into the major extracorporeal circuit by a roller pump. $\dot{Q}$br(s-p) flowed by gravity from the left atrium to the calibrated cylinder where $\dot{Q}$br(s-p) was continuously measured.

Study Protocol and Statistical Analysis

Patients were randomly assigned to groups A and B.

In both groups with the $\dot{Q}$br(s-p) measuring circuit at work 10 minutes were allowed to reduce to ~ 0 cm H_2O the pressure in the aortic root increased by the cardioplegic solution injection ($\sim$700 cc). Indeed for a short time after the cardio-plegic injection some cardioplegic solution flows through the coronary artery and a small amount may reach the left atrium via the Thebesian veins, i.e. it may mix with $\dot{Q}$br(s-p). In group A, $\dot{Q}$br(s-p) was measured for two periods of 20 minutes with alveolar pressure $(P_A) = 4.8 \pm 0.5$ cm H_2O (mean $\pm$ SE). In group B, $\dot{Q}$br(s-p) was measured for 20 minutes with $P_A = 4.6 \pm 0.3$ cm H_2O (Period I).

P_A was then increased to 14.7 ± 1.0 cm H_2O and $\dot{Q}$br(s-p) was measured for further 20 minutes (Period II). In both groups and in each period $\dot{Q}$br(s-p) was calculated as % of cardiac output (pump flow). Data, as in the animal studies, are reported as % changes in relation to control (Period I).

Results

In group A, P_A was constant and Qbr(s-p)also remained constant (Fig. 2). In group B, the increase in P to 14.7 ± 1.0 cm H_2O was associated with a $\dot{Q}$br(s-p) reduction of 39 % (p < 0.02, Fig. 2).

In both groups temperatures, systemic blood pressure and cardiac output presented only minor changes (Table I).

Table I. Hemodynamic parameters and temperatures in patients during total cardio-pulmonary bypass

| | Group A | | Group B | |
	Period I	Period II	Period I	Period II
T es. (°C)	26.8±0.5	26.9±0.2	27.0±0.2	27.3±0.3
T rect. (°C)	29.4±0.5	28.6±0.5	28.8±0.3	28.8±0.3
CO (1/min.)	2.8±0.2	2.6±0.3	2.7±0.2	2.7±0.3
mBP (mmHg)	63.0±4	72.0±5	63.0±4	66.0±3

Human studies. T es. = Esophageal temperature; T rect. = rectal temperature; CO = cardiac output (pump flow); mBP = mean systemic blood pressure.

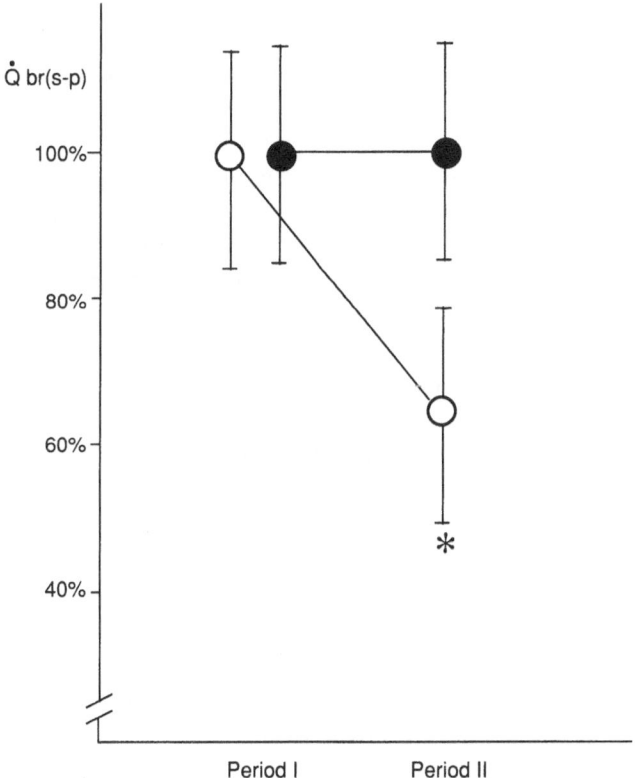

Fig. 2. Human studies. $\dot{Q}$br(s-p) changes (% in relation to period I). Closed circles = group A; Open circles = group B.

Discussion

Our study in dogs confirms previous observations that $\dot{Q}br(s\text{-}p)$ is reduced by PEEP also in a carefully temperature controlled preparation. This observation is relevant because previously the temperature setting of the experimental preparation, at least under some experimental conditions masked, the regulation of $\dot{Q}br(s\text{-}p)$.[5]

The study in humans was necessarily performed in very peculiar and unphysiologic circumstances. However, the data were similar to those obtained in dogs and therefore strengthen the experimental observation.

The temperature setting of the experimental preparation is very important in studies dealing with bronchial blood flow because one of the major tasks of bronchial circulation is the control, through warming and cooling of inspiratory and expiratory gases, of body temperature. This issue is also complicated by the influences of the humidity of respiratory gasses on bronchial circulation.[14]

Both these observations are particularly relevant in the dog because the greatest heat exchange of the dog is through respiration (panting). In has recently been shown that small changes in LLL temperature from canine body temperature markedly increase or reduce $\dot{Q}br(s\text{-}p)$.[3]

What is more, it has been shown that the net $\dot{Q}br(s\text{-}p)$ response to acute pulmonary artery obstruction depends on the temperature setting of the experimental preparation.[5] However, the effect of PEEP on bronchial blood flow does not seem to be temperature dependent because our results are superimposable on those previously obtained in non-temperature controlled preparations.[2,12]

This observation is relevant for the meaning of the studies in humans where body temperature is necessarily low.

The $\dot{Q}br(s\text{-}p)$ measuring techniques in animals and humans differ because, in animals, lung weight changes, used as an indicator of lung fluid changes, are included in the calculation of $\dot{Q}br(s\text{-}p)$. It is not possible to measure lung weight changes in humans during total cardio-pulmonary bypass. This may be relevant to our study in humans because some bronchial blood flow may be pooled in the lung particularly with elevated P_A.

In other words, the effect of PEEP may be to redistribute lung fluids[15] and not to reduce $\dot{Q}br(s\text{-}p)$. However, this study in humans has been performed with the lungs in Zone I ($P_A >$ pulmonary vascular pressure).

It has been demonstrated with *in vivo* experiments that lung weight is constant in Zone I during P_A changes similar to ours.[12] Therefore, our measurements of $\dot{Q}br(s\text{-}p)$ represent the entire $\dot{Q}br(s\text{-}p)$.

The observation that $\dot{Q}br(s\text{-}p)$ is reduced by increasing P_A and therefore by PEEP suggests than the lowest P_A for adequate systemic blood oxygenation should be used to preserve bronchial blood flow during assisted ventilation.

References

1. Charan N.: The bronchial circulation system: structure, function, and importance. Resp. Care 1984; 29: 1226-1235
2. Baile E.M., Albert R.K., Kirk W., Laskminarayan S., Wiggs B.J.R., Parè P.D.: Positive end-expiratory pressure decreases bronchial blood flow in dog. J. Appl: Physiol. 1984; 56: 1289-1293
3. Agostoni P.G., Deffabach M.E., Kirk W., Brengelmann G.L.: Temperature dependence of intraparenchymal bronchial blood flow. Respir. Physiol. 1987; 68: 259-267
4. Magno M., Charan N., Parson G., Baile L., Albertine K., Butler J.: Nomenclature for the bronchial circulation. J. Appl. Physiol. 1987; 62: 2512
5. Agostoni P.G., Deffabach M.E., Kirk W., Laskminarayan S., Mendenhall J.M., Albert R.K., Butler J.: Temperatures alters bronchial blood flow response to pulmonary artery obstruction. J. Appl. Physiol. 1987; 62:1907-1911
6. Agostoni P.G., Deffabach M.E., Kirk W., Laskminarayan S., Butler J.: Upstream pressure for systemic to pulmonary flow from the bronchial circulation in dogs. J. Appl. Physiol. 1987; 63: 485-491
7. Baile E.M., Ling H., Heyworth J.R., Hogg J.C., Paré P.D.: Bronchopulmonary anastomotic and non-coronary collateral blood flow in humans during cardio-pulmonary bypass. Chest 1985; 87:749-754
8. Laskminarayan S., Jindal S.K., Kirk W., Butler J.: Acute increase in anastomic bronchial systemic to pulmonary blood flow due to generalized lung injury. J. Appl. Physiol. 1987: 62: 2358-2361
9. Dalen J., Haffajee C., Alpert J., Howe J., Ockene I., Paraskos J.: Pulmonary embolism, pulmonary hemorrhage and pulmonary infarction. New Engl. J. Med. 1977; 296: 1431-1435
10. Chait A., Summers D., Kransnow N., Wechsler B.M.: Observation of the fate of large pulmonary emboli. Am. J. Roentgenol. Radium Ther. Nucl. Med. 1967; 100: 364-373
11. Cassidy S.S., Haynes M.S.: The effects of ventilation with positive end-expiratory pressure on the bronchial circulation. Respir. Physiol. 1987; 66: 269-278
12. Charan N.B., Albert R.K., Laskminarayan S., Kirk W., Butler J.: Factors affecting bronchial blood flow through bronchopulmonary anastomoses in dogs. Am. Rev. Respir. Dis. 1986; 134: 85-88
13. Wagner E.M., Mitzner W.A., Bleeker E.R.: Effects of airway pressure on bronchial blood flow. J. Appl. Physiol. 1987; 62: 561-566
14. Baile E.M., Guillemi S., Parè P.D.: Tracheobronchial and upper airway blood flow in dogs during thermally induced panting. J. Appl. Physiol. 1987; 63: 2240-2246
15. Gee M. H., Williams D.O.: Effect of lung inflation on perivascular cuff fluid volume in isolated dog lung lobes. Micr. Res. 1979; 17: 192-201

10. Hyperinflation due to Tonic Activity in Inspiratory Muscles

Y. JAMMES, M. BADIER
Respiratory Physiology Laboratory, CNRS, Faculty of Medicine, Marseille, France

Introduction

Tonic activity in inspiratory muscles has been recorded in particular circumstances. This was observed not only in patients suffering from pulmonary diseases but also in healthy humans or animals.

The tonic discharge is characterized by a persistent firing of muscle motor units or motor nerves during expiration. In most cases there are no tonic and sustained contractions throughout the ventilatory cycle because inspiratory modulation persists with incremental myopotentials and muscle force. We will examine the different circumstances where this phenomenon has been described, and then its reflex pathways.

Circumstances

Tonic Inspiratory Activity Associated with Pulmonary Hyperinflation

In humans increased thoracic gas volume is well known during acute bronchospasm produced by inhalation of aerosolized histamine[19] as well as in asthmatic patients.[22] These changes are reversible by administration of bronchodilator substances. More recently, persistent inspiratory muscle contraction throughout expiration was assessed from the observation that the most positive expiratory pleural pressure was less than predicted chest wall relaxation pressure.[9,11]

This indicated that increases in functional residual capacity (FRC) did not parallel the magnitude of intrapulmonary gas trapping due to the bronchoconstriction. EMG recordings in external intercostal muscles (surface electrodes) and the diaphragm (oesophageal electrodes) revealed tonic myoelectrical activity during the hyperinflation associated with histamine-induced bronchospasm in normal or asthmatic subjects.[13] Tonic discharge paralleled the changes in thoracic gas volume (Fig. 1).

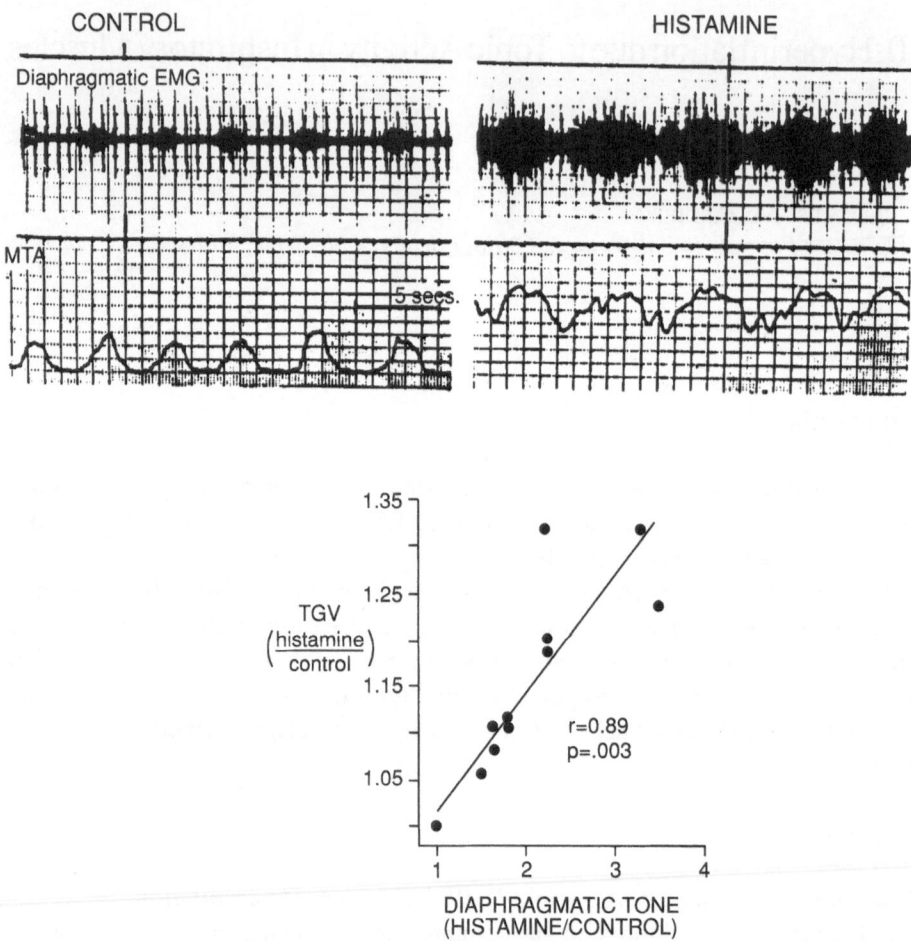

Fig. 1. Tonic activity in diaphragmatic EMG during histamine-induced hyperinflation (top) (Raw and integrated EMG recorded from an oesophageal electrode) and correlation between increase in tonic EMG activity and change in thoracic gas volume (TGV) produced by different doses of histamine (bottom) in normal and ashmatic subjects.[13]

Surprisingly, very few animal studies have been performed in order to demonstrate the existence of tonic inspiratory contractions despite the numerous observations on changes in lung mechanics and also in breathing pattern and ventilatory control in experimental models of asthma. In a recent work, Palevsky et al.[14] re-

ported tonic phrenic activity measured during expiration during acute bronchoconstriction produced by inhalation of Ascaris suum antigen in sensitized dogs. In
rabbits we recorded tonic EMG discharge in the diaphragm and external intercostal
muscles and also in phrenic motoneurons when acute bronchoconstriction with

carbachol

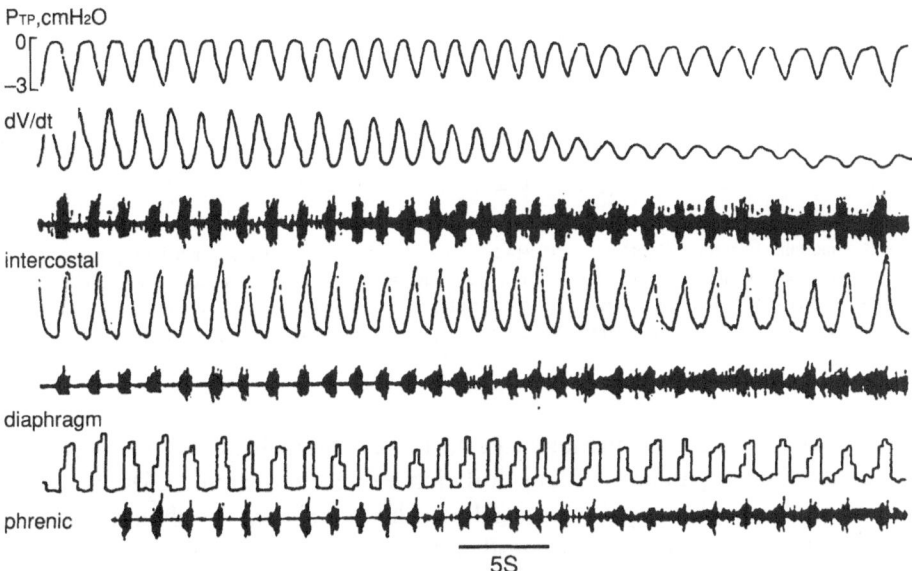

Fig. 2. Tonic activity in external intercostal and diaphragmatic EMG and phrenic motoneurons after i.v. injection
of carbachol (50 mcg) in a rabbit (Transpulmonary pressure:P_{TP}; airflow : dV:dt) raw and intergrated EMGs and
phrenic neurogram. [1]

associated increased thoracic gas volume was produced by i.v. injection of
carbachol (Fig. 2), histamine or phenyldiguanide (PDG).[1] Other observations in
cats breathing dense gas mixtures in hyperbaric conditions, a circumstance leading
to internal mechanical load, have also shown tonic EMG discharge in the diaphragm
and chest wall muscles (Fig. 3).[4]

Tonic Inspiratory Activity without Pulmonary Hyperinflation

This has been demonstrated in anesthetized animals breathing against external
inspiratory (I) loads which produced a progressive decrease in thoracic gas volume.
In cats, resistive or elastic I loads only prolonged the inspiratory discharge in
external intercostal muscles and the diaphragm;[12,17] and also in ventral respiratory
group neurons.[16] However, there was a large discrepancy between results obtained

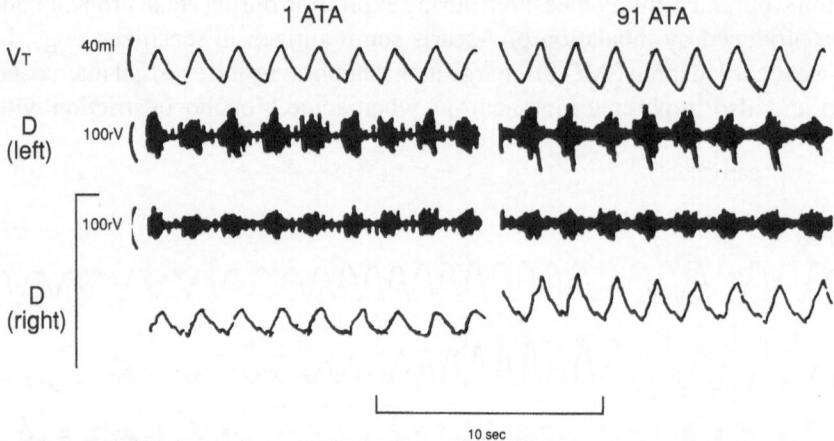

Fig. 3. Simultaneous recordings of tidal volume (Vt) measured with a volume-displacement body plethysmograph and diaphragmatic EMG (right and left cupola) (raw and integrated) in an awake cat at sea level (1 ATA) then during a sojourn at 91 ATA (900 msw) where the density of inspired gas mixture (Helium-Oxygen) was 19 g/l[1]. Absence of complete diaphragmatic relaxation during expiration was observed at maximal pressure.[4]

in different species; then, in rabbits the decrease in thoracic gas volume associated with I loading or with passive lung deflation produced true tonic EMG activity in the diaphragm and external intercostal muscles and also in the phrenic motoneurons[1,15] (Fig. 4a). The level of anesthesia and mostly the choice of anesthetics determined also the magnitude of the prolongation of the inspiratory discharge in response to I loading. This was demonstrated in cats where the lengthening of diaphragmatic contraction in response to tracheal occlusion (TO) was multiplied by 4 when using chloralose instead of pentobarbitone.[6]

Hyperinflation Without Tonic Inspiratory Activity

Breathing against an external expiratory load (resistive or threshold load) increased the thoracic gas volume but shortened the duration of diaphragmatic contraction and also the postinspiratory the fact thatEMG discharge. This was observed in man,[2,10] and also in dogs,[2,5] cats[12] and rabbits.[1] This occurred despite the fact that EMG activity in expiration could be recorded in inspiratory intercostal and scalene muscles in humans breathing against expiratory resistors.[10]

Mechanisms

Sensory Information from the Lungs

Kelsen et al.[8] have compared in the same individuals (normal or asthmatics) the responses to external (mechanical) and internal (bronchospasm) loadings. They showed that facilitatory influences on inspiratory muscles and also increased FRC level were more pronounced during induced bronchospasm than when subjects breathed against external resistive loads. These authors seem to have suggested first that increased sensory inputs from the lungs and airways may explain the different responses. Indeed, most animal observations revealed that vagotomy abolished the tonic inspiratory activity of respiratory muscles; this was assessed in dogs,[14] cats[6] and rabbits.[1,15]

Somes results suggest that thin sensory vagal fibres play a major role in this response. In rats with experimental pneumonia induced by paraquat,[21] the observed increase in functional residual capacity was not reduced during moderate vagal cooling at 8°C, which abolished the Breur-Hering reflex. In contrast, the increase in FRC was reduced after bivagotomy. This suggested first that increased thoracic gas volume was reflexly initiated and second that vagal pulmonary stretch receptors (PSR) did not mediate it.

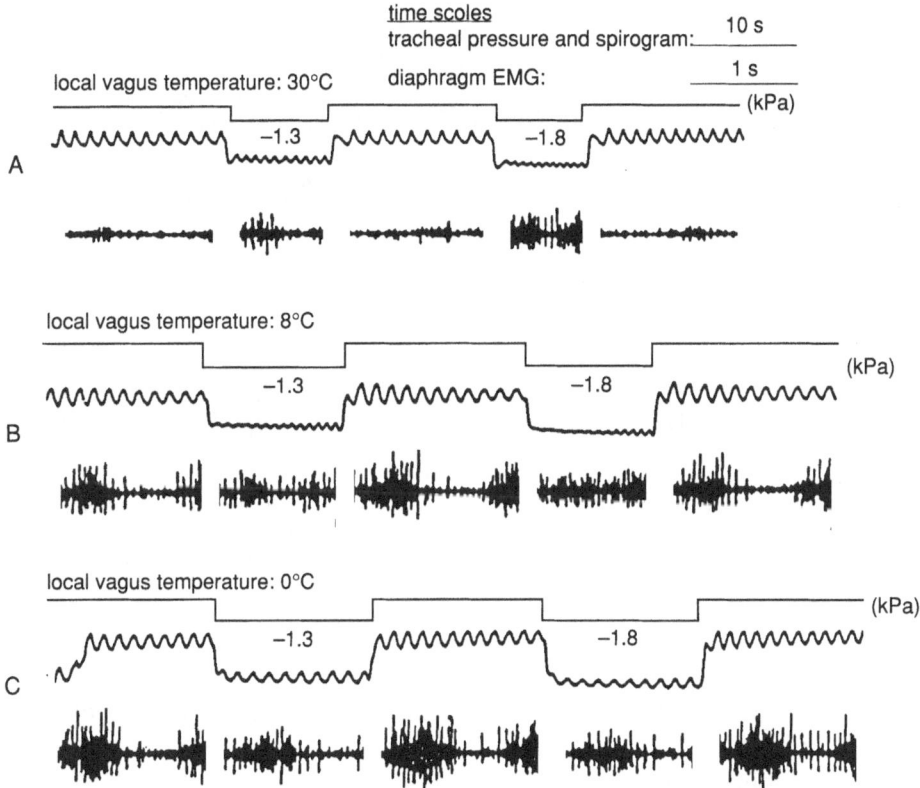

Fig. 4. Tonic diaphragmatic EMG activity induced by lowering the tracheal pressure in a rabbit breathing spontaneously (A). This effect persisted when the vagus nerves were cooled at 8°C (B) but disappeared when the local nerve temperature was 0°C (C).[15]

Another observation in rabbits by Patberg[15] also showed that vagal cooling at 8°C did not abolish but enhanced tonic inspiratory activity in response to lung deflation (Fig. 4b); this effect disappeared only after vagal cooling at 0°C, i.e. in a condition where conduction velocity was blocked in all vagal fibres (Fig. 4c).

In a recent work on rabbits we demonstrated that the tonic EMG discharge in external intercostal muscles and the diaphragm in response to i.v. injection of phenyl-diguanide occurred during procaine block of conduction in thin vagal fibres (Fig. 5).

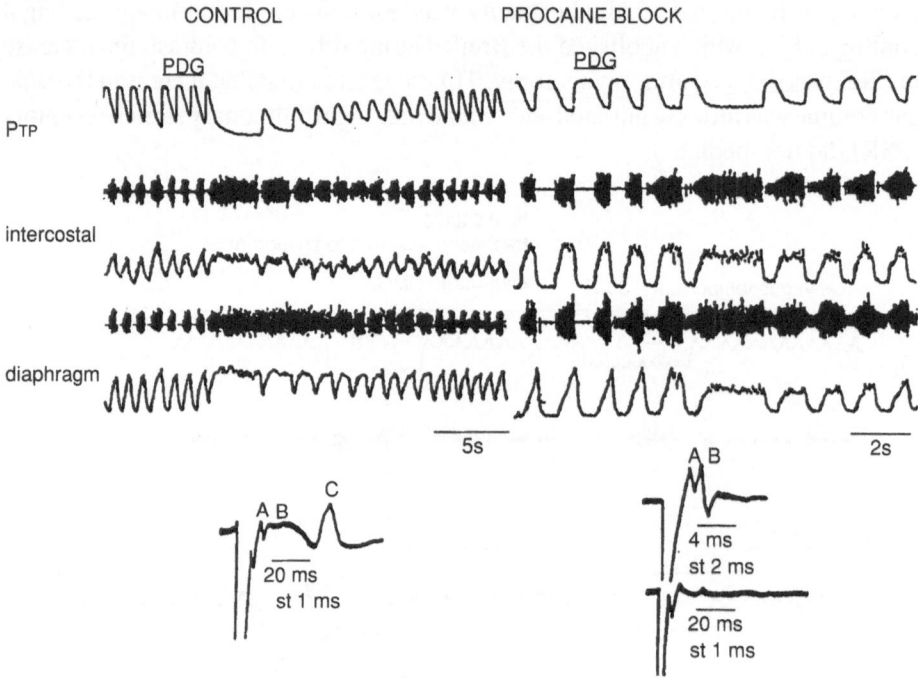

Fig. 5. Tonic contraction of external intercostal muscle and the diaphragm following i.v. injection of phenyldiguanide (PDG, 100 mcg) in a rabbit breathing spontaneously. In control conditions the compound vagal potential (VP) evoked by nerve stimulation shows the A, B and C waves corresponding to large and small myelinated fibres and unmyelinated fibres, respectively. Procaine block of conduction in vagal C fibres (disappearance of the C wave) markedly reduces the tonic EMG response to PDG.

The existence of a bronchoconstriction mediated directly through cholinergic agonists or reflexly via the stimulation of pulmonary vagal afferents is not necessary for the induction of tonic activity in inspiratory muscles.[1]

Indeed, atropine i.v. abolished carbachol-induced bronchospasm and also markedly reduced the reflex bronchoconstrictor component in response to histamine or PDG. In the first circumstance, atropine suppressed the tonic inspiratory activities associated with carbachol-induced bronchospasm. However, in the cases of reflex bronchoconstriction, histamine or PDG continued to evoke tonic EMG discharges despite the fact that the associated bronchospasm was weak or absent after atropine administration.

All these observations converge to show that tonic activity in diaphragm and intercostal muscles results mostly from the stimulation of vagal lung receptors connected to thin afferent fibres.

This may be elicited via the contraction of airway smooth muscle,[20] by decreased lung volumes or through the direct action of PDG and histamine or other inflammatory mediators) on free nerve endings.[3]

Other Reflex Pathways

In particular circumstances, such as infinite inspiratory loading produced by occlusion of the inspiratory line (TO) and also in some species, chest wall and phrenic afferents may participate in the prolongation of the firing duration in dorsal (DRG) and ventral respiratory group neurons (VRG).

This was demonstrated in decerebrate or anesthetized and vagotomized cats by Shannon et al.[17-18] who found that the activation of either phrenic or intercostal muscle afferents by TO prolonged the firing time of inspiratory cells in VRG and DRG and sometimes induced a sustained firing during expiration. However, in the same species other observations led to the conclusion that the inspiratory response TO was reduced but not suppressed after spinal section at C8 level and disappeared only after bivagotomy.[6]

Potential Implications

The absence of complete relaxation of inspiratory muscles during expiration must induce some disturbances in their length-tension relationships, leading to increased sensory pathways from muscle proprioceptors, such as spindles in intercostal muscles and Golgi tendon organs in the diaphragm.[7]

All these conditions may converge to enhance the "sense of breathing" and in consequence the dyspnea during asthma attack and other circumstances where vagal lung receptors are stimulated.

This phenomenon of tonic inspiratory contractions is underestimated because EMG recordings of respiratory muscles are rarely performed not only in asthmatic patients but also in animal models of asthma.

82

References

1. Badier M., Jammes Y., Romero-Colomer P., Lemerre C.: Tonic activity evoked in inspiratory muscles and phrenic motoneurons by stimulation of vagal afferents. J. Appl. Physiol. 1989; 66 (4):1613-1619
2. Campbell E.J.M., Dichinson C.I., Dinnick O.P., Howell J.B.L.: The immediate effects of threshold loads on the breathing of men and dogs. Clin. Sci. 1961; 21 : 309-320
3. Coleridge H.M., Coleridge J.C.G.: Reflexes evoked from tracheobronchial tree and lungs. In : Fishman A.P. (Ed.) *Handbook of Physiology*, Section. 3 : *The Respiratory System*, vol. 2, part 1. Am. Physiol. Society, Bethesda, 1986, pp. 395-429
4. Imbert G., Jammes Y., Naraki N., Duflot J.C., Grimaud C.: Ventilation, pattern of breathing and activity of respiratory muscles in awake cats during oxygen-helium simulated dives (1000 msw). In: Bachrach A.J., Matzen M.M. (Eds.), *Underwater Physiology*, Tome VII, Undersea Medical Society, Bethesda, 1981, pp. 273-282
5. Jammes Y., Bye P.T. B., Pardy R.L., Katsardis C., Esau S., Roussos C.: Expiratory threshold load under extracorporeal circulation : effects of vagal afferents. J. Appl. Physiol., 1983; 55:307-315
6. Jammes Y., Mathiot M.J., Delpierre S., Grimaud C.: Role of vagal and spinal sensory pathways on eupneic diaphragmatic activity. J. Appl. Physiol. 1986; 60 : 479-489
7. Jammes Y.: Cesthwall and diaphragmatic afferents : their role during external mechanical loading and respiratory muscle ischemia. In : Grassino A., Fracchia C., Rampulla C., Zocchi L. (Eds.): *Respiratory Muscles in Chronic Obtructive Pulmonary Disease*, London, Springer Verlag, Verona Bi & Gi Editori, 1988, pp. 49-57
8. Kelsen S.G., Prestel T.F., Cherniack N.S., Chester E.H.,Deal C.: Comparison of the respiratory responses to external resistive loading and bronchoconstriction. J. Clin. Invest. 1981; 67 : 1761-1768
9. Martin J.C.,Powell E., Shore S., Emrich J., Engel L.A.: The role of respiratory muscles in the hyperinflation of bronchial asthma. Am. Rev. Respir. Dis. 1980; 121, 441-447
10. Martin J.G., Habib M.,Engel L.A.: Inspiratory muscle activity during induced hyperinflation. Respir. Physiol. 1980; 303-313
11. Martin J.G., Shore S.A., Engel L.A.: Mechanical load and inspiratory muscle action during iduced asthma. Am. Rev. Respir. Dis. 1983; 128 : 455-460
12. Mathiot M.J.,Jammes Y.,Grimaud C.: Role of vagal and spinal sensory pathways in diaphragmatic response to resistive loading. Neuroscience Letters 1987; 73 : 131-136
13. Muller N., Brian A.C., Zamel N.: Tonic inspiratory muscle activity as a cause of hyperinflation in histamine-induced asthma. J. Appl. Physiol. 1980; 49 : 869-874
14. Palevsky H.I., Grippi M.A., Pack A.I.: The effect of antigen-induced bronchoconstriction on phrenic nerve activity. Am. Rev. Respir. Dis. 1986; 133 : 749-756
15. Patberg W.R.: Effect of graded vagal blockade and pulmonary volume on tonic inspiratory activity in rabbits. Pflügers Arch. 1983; 398 : 88-92
16. Seaman R..G., Zechman F.W., Frazier D.T.: Response of ventral respiratory group inspiratory neurons to mechanical loading. J. Appl. Physiol. 1983; 54 : 254-261
17. Shannon R., Zechman F.W.: The reflex and mechanical response of the inspiratory muscles to an increased airflow resistance. Respir. Physiol. 1972; 16 : 51-69
18. Shannon R., Shear W.T., Mercak A.R., Bosler D.C. Lindsey B.G.: Non-vagal reflex effects on medullarly inspiratory neurons during inspiratory loading. Respir. Physiol. 1985; 60 : 193-204
19. Sonne L.M., Georg J.: The respiratory changes during an attack of bronchial asthma. Acta Med. Scand. Suppl. 1950; 239 : 333
20. Vidruk E.H.,Hann H.L., Nadel J.A., Sampson S.R.: Mechanics by which histamine stimulates rapidly adapting receptors in dog lungs. J. Appl. Physiol. 1977; 43 : 397-402
21. Vizek M., Frydrychova M., Houstek S., Palecek F.: Effect of vagal cooling on lung functional residual capacity in rats with pneumonia. Bull. Europ. Physiopath. Resp. 1983; 19 : 23-26
22. Woolcok A.J., Rebuck A.S., Read J.: Lung volume changes in asthma measured concurrently by two methods. Am. Rev. Respir. Dis. 1971; 703-709

11. Effects of Hyperinflation on Respiratory Muscle Function

M. DECRAMER

Respiratory Department, University Hospital, Catholic University, Leuven, Belgium

Introduction

Although the effects of hyperinflation on the expiratory muscle are still poorly understood, it is generally accepted that it adversely affects the function of inspiratory muscle,[1-2] due to shortening of these muscles which places them at a disadvantageous portion of their force-length relationship. Moreover, changes in diaphragmatic geometry,[3] changes in the mechanical arrangement between the two parts of diaphragm,[4] and changes in diaphragm-rib cage interaction occur.[5] Much less, however, is presently known about the effects of hyperinflation on the extradiaphragmatic musculature, which contributes significantly to respiratory acts, even to quiet breathing in man,[6-7] as well as in experimental animals.[8-9]

Improved insight into how hyperinflation changes respiratory muscle function has recently been obtained through the development of animal models of hyperinflation. The experimental evidence obtained in such models and its significance to humans and patients with hyperinflation is reviewed and summarized. Distinction between animal models of chronic hyperinflation and models of acute hyperinflation should be introduced.

Animal Models of Chronic Hyperinflation

From experiments in emphysematous hamsters, it is clear that the diaphragm adapts to chronic hyperinflation.[10-11] The adaption is similar to what occurs in other skeletal muscles and consists of a drop-out of sarcomeres,[12] such that as the in situ operational length shortens, the optimal length shifts to a shorter length as well. As a consequence, the in situ operational length and the optimal length remain matched, and the force generating capacity of the diaphragm is better preserved at the in situ operational length.[13] Although this hamster model of chronic hyperinflation is well developed and the concepts derived from it are well established, it remains questionable whether similar adaptations occur in human emphysema.[14]

Animal Models of Acute Hyperinflation

In any event, the forementioned adaptation does not occur in acute hyperinflation, since there is not sufficient time available. Consequently, inspiratory muscles are definitely expected to operate at a disadvantageous portion of their force-length relationship. We recently examined the effects of acute hyperinflation on inspiratory muscle mechanics in a series of experiments on supine anesthetized dogs.[8,15-17]

First, we measured the changes in length undergone by the diaphragm and the parasternal intercostals during passive inflation from functional residual capacity, FRC, to total lung capacity, TLC, in nine animals using piezo-electric crystals implanted in these muscles.[8,15]

The costal and crural parts of the diaphragm shortened similarly (-33.2 ± 12.6 vs. -30.6 ± 9.7 %, NS), whereas the parasternal intercostals shortened considerably less (7.2 ± 3.2 % 39 interspaces). Assuming that the diaphragm and the parasternal intercostals have a length-tension curve of similar shape, and moreover, that both have their optimal length close to FRC, this observation leads to the conclusion that hyperinflation induces a mechanical disadvantage in the diaphragm which is much more pronounced than in the parasternal intercostals. Subsequent work by Farkas et al.,[18] however, showed that the parasternal length-tension curve was significantly more contracted along its length-axis than the diaphragmatic length-tension curve, but concomitantly that the optimal length of the parasternal intercostals probably occurred at a lung volume above FRC, and thus supported our conclusion.

Second, we examined the effects of hyperinflation on the pattern of thoracicoabdominal motion, pleural and gastric pressure development during inspiration, and EMG activities of the diaphragm and parasternal intercostals, in fourteen vagotomized animals. We found that when animals were made to breathe near TLC by applying continuous positive pressure to the airway opening, they showed inspiratory rib cage expansion frequently associated with abdominal paradox and an inspiratory fall in gastric pressure which were both indicative of ineffective diaphragmatic contraction. Phrenicotomy did not affect this pattern of thoracicoabdominal motion or gastric pressure development, which further demonstrated the ineffectiveness of diaphramatic contraction near TLC (Fig. 1).

Near TLC, EMG activity in the costal part of the diaphragm slightly increased to 125 ± 12 % (p<0.01) of the FRC value and the activity in the crural part of the diaphragm remained relatively constant (107 ± 21 %), while the activity in the

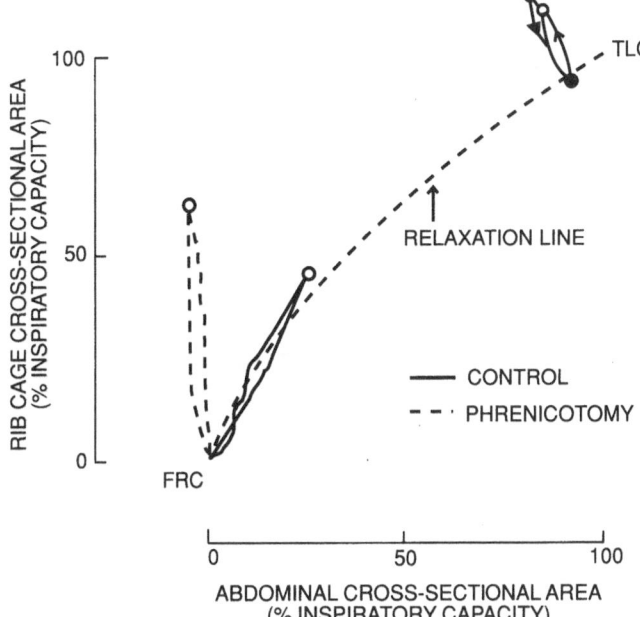

Fig. 1. Konno-Mead diagram showing the relationship between rib cage cross-sectional area and abdominal cross-sectional area in one representative experiment before (solid line) and after (dotted line) bilateral phrenicotomy. Note that at FRC the animal breathes along the relaxation line, while near TLC rib cage expansion associated with abdominal paradox is observed. Phrenicotomy affects the pattern of thoracicoabdominal movement at FRC, but has almost no effect near TLC.[16]

parasternals tended to decrease (75 $\pm$ 27 %, p=0.053). The latter observations showed that the breathing pattern observed could not be explained on the basis of alterations in the pattern of motor unit recruitment.

Surprisingly, the fall in pleural pressure at FRC and near TLC were similar (-6.1 $\pm$ 0.6 vs. -7.5 $\pm$ 1.0 cm H_2O, NS). In view of the only minor alterations in EMG activity observed, this suggested that the pressure generating capacity of the inspiratory muscles was well preserved near TLC. Since diaphragmatic contraction near TLC was shown to be clearly ineffective, other muscles must have compensated for this loss in diaphragmatic effectiveness and their pressure generating capacity should have been better preserved or even improved between FRC and TLC. The parasternal intercostals are known to be other major agonists of inspiration.[6,8] We, therefore, studied the effects of hyperinflation on the mechanical effectiveness of these muscles in a third series of experiments in seven animals.

Force outcome of parasternal intercostal contraction was estimated by measuring changes in intramuscular pressure using a microtransducer method as described previously.[19] In a previous study we demonstrated that the force generated by parasternal intercostal contraction was linearly related to the increase in intramuscular pressure occurring during contraction.[19] The following observations were made in the present study:

1) During direct muscle stimulation the increase in intramuscular pressure for a given supramaximal stimulus tended to be greater at TLC than at FRC (140.7 $\pm$ 28.6

vs. 100 ± 28.3 cm H_2O, NS). The force exerted on the rib was greater at TLC than at FRC (277.4 ± 60.6 vs. 214.2 ± 47.1 g, $p<0.05$).

2) During quiet breathing near TLC, swings in parasternal intramuscular pressure were on the average similar at FRC and near TLC (50.4 ± 16.5 vs. 48.7 ± 13.3 cm H_2O). Since the electrical activity of the parasternals clearly decreased near TLC in this study (82.9 ± 5.5 %, $p<0.01$), a greater pressure development for a given electrical activity was found near TLC (Fig. 2).

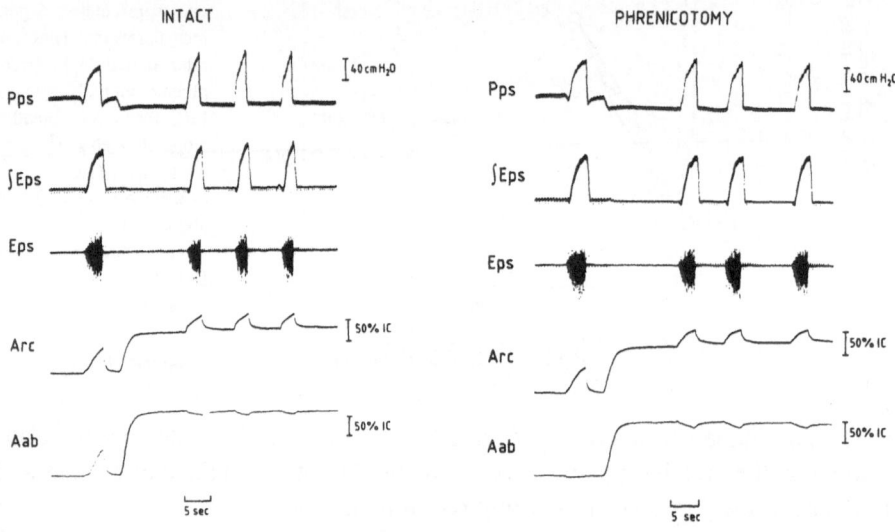

Fig. 2. Tracing obtained in one representative experiment, before (left panel) and after (right panel) bilateral phrenicotomy. In the first breath, the animal breathes at FRC and in the subsequent breaths it breathes near TLC. From top to bottom are parasternal intramuscular pressure (Pps), integrated parasternal EMG (ʃEps), raw EMG (Eps), rib cage cross-sectional area (Arc), and abdominal cross-sectional area (Aab). Note that during inspiration near TLC for a given electrical activation a greater rise in intramuscular pressure is observed.

These two observations indicated that for the parasternal intercostals, the gain converting electrical activity into force development tended to be greater near TLC than at FRC. This suggests, in line with the predictions made by Farkas et al.,[18] that the optimal length of the parasternal intercostals was closer to TLC than to FRC, and consequently that in contrast to its detrimental effect on the diaphragm, hyperinflation was far less detrimental and probably even beneficial to the parasternal intercostals.

Limitations of These Animal Models

It should be emphasized that the analysis developed above has at least three important limitations. First, it is always a relatively difficult process to extrapolate these data obtained in experimental animals to humans and patients with hyperinflation. Nevertheless, some obvious similarities are present between the changes in inspiratory muscle mechanics induced by hyperinflation in these animals and the changes observed in normal subjects and patients.

Second, so far we have only studied the effects of hyperinflation on two muscles, the diaphragm and the parasternal intercostals. Evidently, more muscles could contribute to respiration, including the scalenes,[7,20] sternocleidomastoids,[20] levatores costae,[21] triangularis sterni,[9] and transversus abdominis.[22]

Third, we have mainly concentrated on the changes in force-length characteristics by hyperinflation, while the force generating capacity of the inspiratory muscles could be affected by other factors such as muscle geometry and mechanical arrangement among muscles as well.

Significance for Humans and Patients with Hyperinflation

Although extrapolation of data obtained in animal models to humans and patients is always problematic, several lines of evidence suggest that similar concepts presumably apply to humans and to patients with hyperinflation.

First, although maximal inspiratory pressure decreases sharply with increasing lung volume,[23] the pressure generated by the inspiratory muscles as given by the difference between maximal inspiratory pressure and the elastic recoil curve of the total respiratory system is clearly more independent of lung volume.[24] This suggests that the pressure generating capacity of the inspiratory muscles remains relatively constant with hyperinflation (up to about 80% of the vital capacity) in humans as well.

Second, the pattern of chest wall motion and inspiratory gastric pressure development in normal subjects breathing above FRC is very similar to what we observed in animals.[25-26] Finally, also patients with hyperinflation exhibit signs of diaphragmatic ineffectiveness, the latter being related to the degree of air flow obstruction.[27-29]

Conclusion

Hyperinflation profoundly affects respiratory muscle interaction in dogs, leading to an increased contribution of the parasternal intercostals to breathing. This is explained on the basis of a greatly impaired mechanical effectiveness of the diaphragm, but an improved mechanical effectiveness of the parasternals. Evidence is present that similar concepts presumably apply in humans and patients with hyperinflation. On this basis the relevance of diaphragmatic breathing exercises in patients with hyperinflation is questionable.

References

1. Decramer M., Demedts M., Rochette F., Billiet L.: Maximal transrespiratory pressures in obstructive lung disease. Bull. Eur. Physiopathol. Respir. 1980; 16:479-490

2. Rochester D.F., Braun N.M.T.: Determinants of maximal inspiratory pressure in chronic obstructive pulmonary disease. Am. Rev. Respir. Dis. 1985; 132:42-47

3. Roussos C., Macklem P.T.: The respiratory muscles. New Engl. J. Med. 1982; 307:786-797.

4. Decramer M., De Troyer A., Kelly S., Macklem P.T.: Mechanical arrangement of costal and crural diaphragms in dogs. J. Appl. Physiol. 1984; 56:1484-1490

5. Zocchi L., Garzaniti N., Newman S., Macklem PT.: Effect of hyperinflation and equalization of abdominal pressure on diaphragmatic action. J. Appl. Physiol. 1987; 62:1655-1664

6. De Troyer A., Sampson M.: Activation of the parasternal intercostals during breathing efforts in human subjects. J. Appl. Physiol. 1982; 52:524-529

7. De Troyer A., Estenne M.: Coordination between rib cage muscles and diaphragm during quiet breathing in humans. J. Appl. Physiol. 1984; 57:899-906

8. Decramer M., De Troyer A.:Respiratory changes in parasternal intercostal length. J. Appl. Physiol. 1984; 57:1254-1260

9. De Troyer A., Ninane V.: The triangularis sterni: a primary muscle of breathing in the dog. J. Appl. Physiol. 1986; 60:14-21

10. Supinsky G.S., Kelsen S.G.: Effect of elastase-induced emphysema on the force generating of the diaphragm. J. Clin. Invest. 1982; 70:978-988

11. Farkas G.A., Roussos C.: Adaptability of the hamster diaphragm to exercise and/or emphysema. J. Appl. Physiol. 1982; 53:1263-1272

12. Farkas G.A., Roussos C.: Diaphragm in emphysematous hamsters: sarcomere adaptability. J. Appl. Physiol. 1983; 54:1635-1640

13. Oliven A., Supinsky G.S., Kelsen S.G.: Functional adaptation of diaphragm to chronic hyperinflation in emphysematous hamsters. J. Appl. Physiol. 1986; 60:225-231

14. Arora N.S., Rochester D.F.: COPD and human diaphragm muscle dimensions. Chest 1987; 91:719-724

15. Decramer M., Jiang T.X., Reid M.B., Kelly S., Macklem P.T., Demedts M.: Relationship between diaphragm length and abdominal dimensions. J. Appl. Physiol. 1986; 61:1815-1820

16. Decramer M., Jiang T.X., Demedts M.: Effects of acute hyperinflation on inspiratory muscle. J. Appl. Physiol. 1987; 63:1493-1498

17. Jiang T.X., De Schepper K., Demedts M., Decramer M.: Effects of acute hyperinflation on the mechanical effectiveness of the parasternal intercostals. Am. Rev. Respir. Dis. 1989; 139 (2): 522-528

18. Farkas G.A., Decramer M., Rochester D.F., De Troyer A.: Contractile properties of intercostal muscles and their functional significance. J. Appl. Physiol. 1985; 59:528-535

19. Leenaerts P., Demedts M., Decramer M.: Respiratory changes in parasternal intercostal intramuscular pressure. (abstract). Fed. Proc. 1987; 46:819

20. Farkas G.A., Rochester D.F.: Contractile characteristics and operating lengths of canine neck inspiratory muscles. J. Appl. Physiol. 1986; 61:220-226

21. Goldman M.D., Loh L., Sears T.A.: The respiratory activity of human levator costae muscles and its modification by posture. J. Physiol. (Lond.) 1985; 362:189-204

22. Gilmartin J., Ninane V., De Troyer A.: Abdominal muscle use during breathing in the anesthetized dog. Respir. Physiol. 1987; 70:159-171

23. Rinqvist T.: The ventilatory capacity in healthy subjects. Scand. J. Clin. Lab. Invest. 1966; 18(Supplement 88):1-113

24. Rahn H., Otis A.B., Chadwick C.A, Fenn W.: The pressure volume diagram of the thorax and the lung. Am. J. Physiol. 1946; 146:161-178

25. Camus P., Desmeules M.J.: Chest wall movements and breathing pattern at different lung volumes (abstract). Chest 1982; 243

26. Wolfson D.A., Strohl K.P., Dimarco A.F., Altose M.D.: Effects of an increase in end-expiratory volume on the pattern of thoracoabdominal movement. Respir. Physiol. 1983; 53:273-283

27. Sharp J.T., Goldberg N.B., Druz W.S., Fishman H.C., Danon J.: Thoracoabdominal motion in chronic obstructive pulmonary disease. Am. Rev. Respir. Dis. 1977; 115:47-56

28. Gilmartin J.J., Gibson G.J.: Abnormalities of chest wall motion in patients with chronic airflow obstruction. Thorax 1984; 39:264-271

29. Gilmartin J.J., Gibson G.J.: Mechanisms of paradoxical rib cage motion in patients with obstructive pulmonary disease. Am. Rev. Respir. Dis. 1986; 134: 683-687

12. The Effect of Lung Hyperinflation on Inspiratory Muscle Function

A. GRASSINO, P. BEGIN
Notre Dame Hospital and Meakins-Christie Laboratories McGill University, Montreal, Canada

Introduction

There are very good reasons to believe that lung hyperinflation plays a crucial role in the deterioration of inspiratory muscle function. In fact, it is very likely that chronic respiratory failure (hypoxemia and hypercapnia) in chronic obstructive pulmonary disease (COPD patients may develop because of the mechanical disadvantages of the inspiratory muscles rather than (or in addition to) disruption of the ventilation perfusion ratio. The mechanisms of ventilatory failure in COPD are complex and through the years have captured the attention of many physiologists and clinicians who have contributed in-depth analyses of some of the possible mechanisms leading to chronic failure. We review next some of the evidence available in the literature of the last 25 years, and we will show how the interpretations of chronic respiratory failure mechanisms evolved.

Fighter vs Non-fighter Theory

It was the clinicians' acute sense of observation that led to the description of two types of COPD patients. The blue bloaters and the pink puffers. Such grouping was made mainly by the observation that some patients become hypoxemic and hypercapnic with a history of periods of chronic failure (hypoxia-hypercapnia) over the years, while another group, equally obstructive and dyspneic, do maintain homeostasis and keep the blood gases at near normal values during most of their lives. It is only at the end of the evolution of this disease that they fail in their fight and die of respiratory failure as the blue bloaters do.

In the fifties physiologists thought that there should be a difference in the response of the respiratory center in the two types of patients to $PaCO_2$ or PaO_2. A failure of neurological origin in blue bloaters was postulated as the cause. Robin and O'Neill[1] developed the appealing label of fighters and non-fighters to these two groups of COPD patients. It is a graphical description which classifies the patients and offers as a mechanism no more than an intuitive hypothesis, which is quite reasonable, but unproven. After all, there was circumstantial evidence that the respiratory center can fail as in the Ondine curse syndrome or even in the Pickwickian type of obese patients. Non-fighters were called "lazy" breathers, because if they were forced into voluntary hyperventilation they could reduce the $PaCO_2$.

The Breathing Pattern Theory

In the early 1970's there was a surge of interest in neuromechanical control of breathing, spearheaded by an interesting combination of new research on neural control of breathing at the cellular level and at the neuromuscular response level. One of the consequences of this approach was the realization that minute ventilation is a poor parameter for the evaluation of neuromuscular activity because it is strongly affected by airways resistance. A decreased VE can be secondary to lack of central drive or secondary to limitation in air flow due to high airways resistance.

Whitelaw et al.[2] proposed that the mouth occlusion pressure developed during the first 100 ms of an occluded breath could separate neuromuscular drive from lung impedance. As soon as the method was refined for clinical use, Sorli et al.[3] tested the fighter vs non-fighter hypothesis by measuring P.1 in clinically stable COPD with chronic hypercapnia and in normocapnic COPD, both age and sex matched. A group of normal subjects was evaluated as well.

P.1 as well as other parameters related to the breathing cycle, gas exchange and ventilation perfusion were measured in all subjects during resting breathing.

Fig. 1 shows some of the significant findings. The P.1 values in normocapnic and hypercapnic patients were very high in relation to normal subjects, but, were similar in both groups of patients. This was interpreted as a blow to the fighter vs non-fighter theory: here there was real evidence that both groups of COPD were fighters. The study however was not designed to answer why some patients develop hypercapnia, and the analysis from the ventilation parameters did show that hypercapnia was probably caused by a high V_D/V_T, the CO_2 production and ventilation being very similar in both groups.

Sorli et al.[3] speculated that the high V_D/V_T associated with hypercapnia was caused by a rapid shallow breathing pattern which in turn could be caused by irritant reflexes originated in chronic bronchitis which was much more prevalent in hypercapnic than normocapnic COPD patients. They also found that the patients

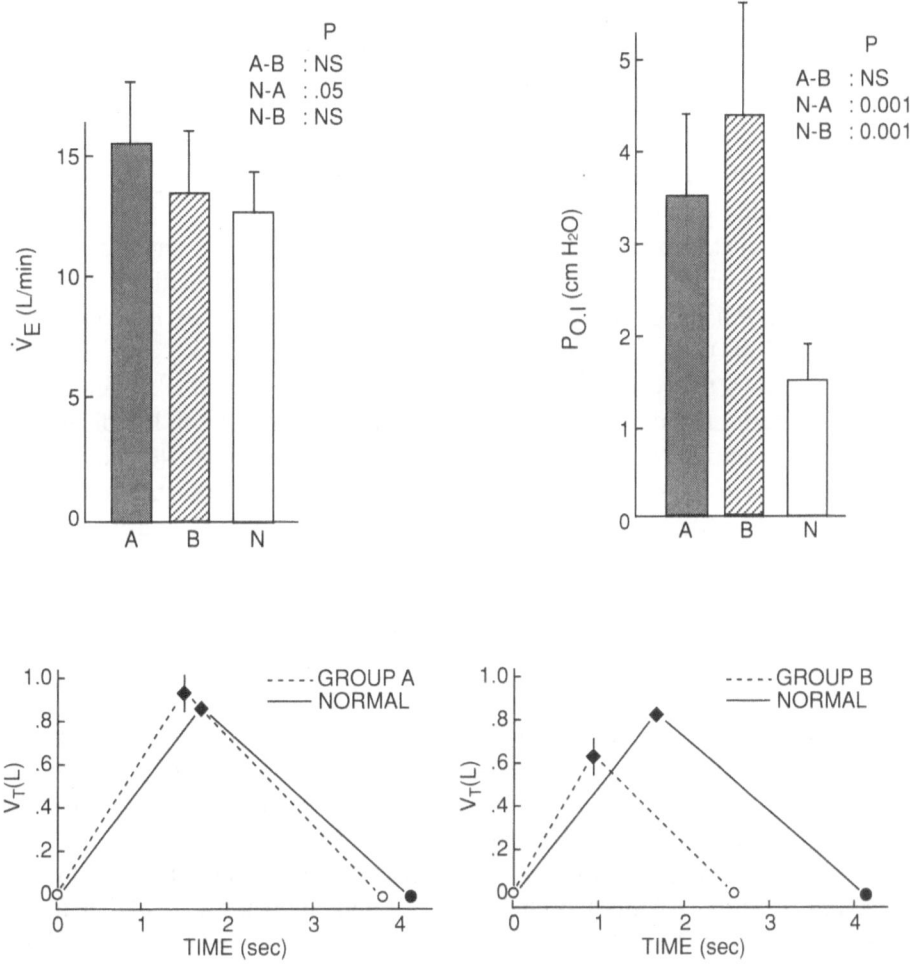

Fig. 1. Upper panels: values of ventilation at rest, VE and mouth occlusion pressure (P.1), in a group of hypercapnic COPD patients (B) normocapnic COPD (A) and normal subjects (N). T test significance by pairs is shown in the right upper corner.

Lower panels: tidal volume of normal subjects contrasted with normocapnic (A) and hypercapnic (B) patients.[3]

in the group were somehow more hyperinflated, and their impedance (measured as $P.1/V_T/T_I$) was higher.

In summary, Sorli et al.[3] thought that COPD patients were all fighters. Some of the patients retained CO_2 because of their fast-shallow breathing pattern hypothetically caused by chronic bronchitis originated reflexes.

The Fatigue Threshold Theory

In the late seventies the concept of respiratory muscle fatigue was developed and research from the Meakins-Christie Laboratories in Montreal led by Dr. P.T. Macklem explored this new fascinating concept and its implications for COPD breathing. At the time it seemed quite logical to think that fatigue may cause respiratory muscle failure and this will lead to hypercapnia.[4]

Bellemare et al.[5,6] explored the mechanisms leading to fatigue and found that, as a first approximation, we can predict if the diaphragm will or will not fatigue by knowing the Pdi developed on each breath and the timing of its contraction-relaxation (T_i/T_T).

Fig. 2 shows this concept validated by experimental results obtained from normal subjects. The dashed line is an iso endurance band of 1 h. It indicates that if the T_i/T_T developed on each breath by the diaphragm is about .40 and the mean Pdi is about .4 of Pdi maximal, this breathing pattern could be sustained for about 1 h. The dashed line is the product of Pdi/Pdimax times T_i/T_T and is called the tension time index. If T_i/T_T increases, the pressure that can be sustained for one hour will be smaller. If a continuous contraction of the diaphragm ($T_i/T_T=1$) will be imposed, fatigue will develop at a pressure of as little as 15% of Pdimax. A short T_i/T_T on the other hand allowed much higher pressures to be maintained for one hour. This figure shows that there is a pressure-time band which describes all the patterns of breathing that the diaphragm can sustain for one hour. A tension time index higher than .15-.18 could not be sustained for one hour. TTdi lower than .15-.18 can presumably be sustained much longer than one hour. All this is valid for the muscle length present at FRC and flows below 1/l/sec.

That being established,[6] the immediate question was where the COPD resting breathing pattern fits in this diagram. 20 COPD, including normo- and hypercapnic subjects had their T_i/T_T and Pdi (expressed as a percentage of maximal Pdi) measured during resting breathing.

Each data point in Fig. 2 represents the average value of each patient during resting breathing. Most of these are much closer to the TTdi one hour endurance threshold than those of normal subjects; however, they are not fatigued.

The next question proposed was will these patients develop fatigue if forced to breathe on the higher side of the threshold?[7] In fact, a group of these patients (indicated by crosses in Fig. 2) were asked to do that by voluntarily controlling their breathing patterns. The patients increased the T_i/T_T and Pdi (breathing slower and deeper) and kept that pattern for several minutes. Fig. 3 shows the results of breathing with the imposed pattern. The patients decreased the H/L ratio of the EMG of the diaphragm (an index of acute fatigue) as soon as the new pattern was adopted; after a few minutes they showed decreased Pdi, and later interrupted the imposed pattern, returning to their spontaneous pattern. By breathing with the

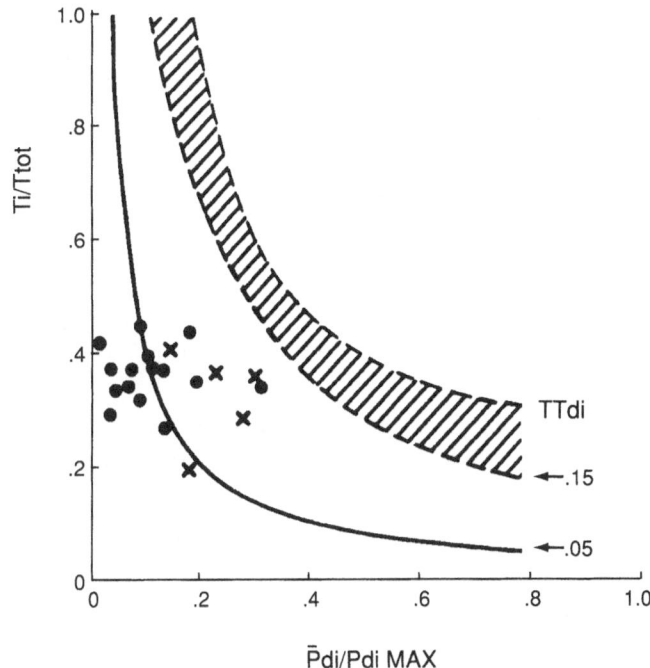

Fig. 2. Relationship between T_I/T_{TOT} and Pdi/Pdi max in 20 patients with COPD while breathing air spontaneously at rest. Each point represents one patient. Crosses indicate the patients in which the pattern of breathing was subsequently modified. TTdi isopleths of .05 and .15 are drawn together with the hatched band describing the fatigue threshold of the diaphragm.[6]

imposed pattern they can hyperventilate, and if hypercapnic they can reduce the $PaCO_2$.

However, if they do so, they will develop fatigue of the diaphragm and would not be able to hold the CO_2 down for very long because of the fall in Pdi.[7] This general behaviour was shown in other experiments in which voluntary hyperventilation was imposed as well.[8-10] Bellemare's experiment did not explore the effect on hypercapnia in particular, but rather concentrated on demonstrating the development of fatigue. We raised the question why do some stable hypercapnic patients while not fatigued during resting breathing become hypercapnic? Could it be that Sorli's finding of insufficient V_T (and smaller Pdi swings), resulting in higher VD/V_T was a mechanism to defend the muscle from fatigue? Would high inspiratory loads work as a negative feedback inducing the respiratory centers to decrease TTdi so as to avoid overload and fatigue? This concept was not further developed at that time.

Evidence Supporting the Concept that Hypercapnia May Develop to Avoid Fatigue

COPD patients have lung hyperinflation as evidenced by a high functional residual capacity/total lung capacity (FRC/TLC) ratio and a decrease in the total

96

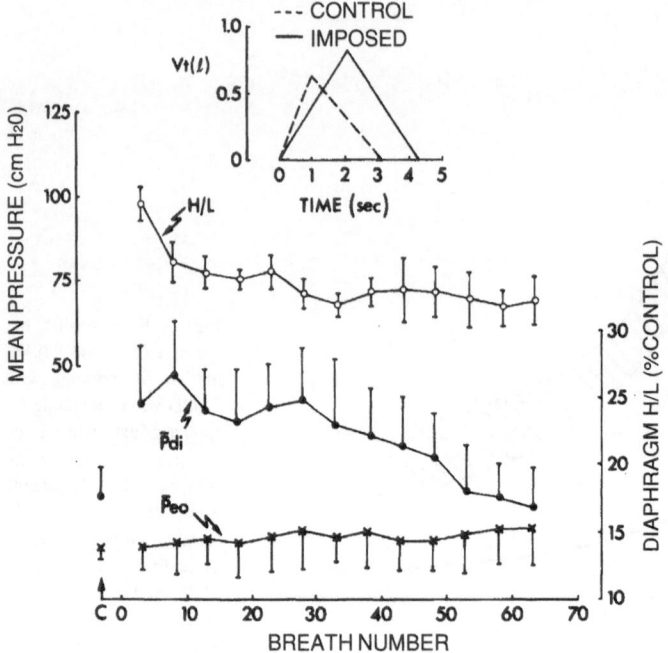

Fig. 3. Effect of the imposed pattern of breathing in five patients with COPD on the time course of diaphram H/L (o) of the mean Pdi (Pdi:) and mean esophageal pressure (Peo:x) during the first 65 breaths. Each point is the mean of five successive breaths. The H/L is expressed as % of the first 3 breaths. The mean gastric pressure is given by the difference between Pdi and Peo. The bars indicate ± 1 SEM. C: control value. At the top of the figure is shown the mean schematic spirogram during control spontaneous breathing and during the imposed pattern of breathing. Vt: tidal volume.[6]

length of the diaphragm as shown by radiographic exam. The values of maximal Pdi are considerably lower.[11] On the other hand, the inspiratory pressure swings are higher than in normal subjects because of the increased lung resistance. It is expected that the Pi/MIP max ratio should be higher, as shown in Fig. 4. This figure shows mean values of maximal inspiratory pressure (MIP) and pressure swings required in resting breathing in normal subjects, in normocapnic and in hypercapnic COPD patients. In the latter group, the peak Pi was closer to or exceeded the fatigue threshold. Few experiments on steady state prolonged inspiratory loads in spontaneously breathing subjects are available.

Grassino and Bellemare reported that upon imposition of a high inspiratory resistive load normal subjects react differently and during the first few minutes some considerable oscillation in breathing pattern occurs, as if the subject was searching for an optimal pattern. Studies by Martyn et al.[12] imposing incremental threshold inspiratory loads for two minute periods show initial hyperventilation (with lower loads), and a progressive return to normocapnia, reaching hypercapnic values when resistance was high enough to be close to the fatigue threshold. Furthermore, a chronic mild airway resistance applied to awake sheep resulted in hypercapnia without evidence of diaphragmatic fatigue.[10]

Recent studies reported by Begin et al.[13] in which over 200 patients had their MIP, esophageal pressure swing, compliance and lung resistance measured during resting breathing indicate some revealing features.

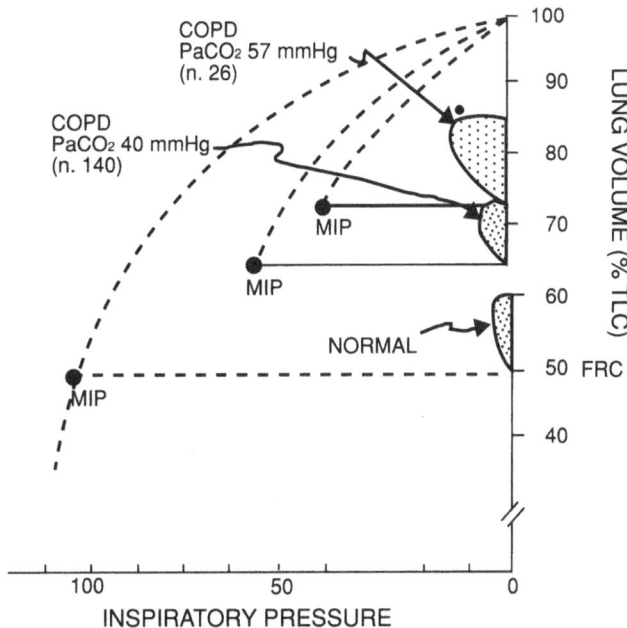

FORCE RESERVE IN
COPD WITH NORMOCAPNIA OR
HYPERCAPNIA

Fig. 4. Maximal inspiratory pressure at the mouth cmH$_2$O is plotted at FRC which expresses as % of TLC. Data is shown for normals, a group of normocapnic COPD patients and a group of hypercapnic COPD patients. The loops are typical pleural pressure swings. The small square block on the MIP lines represent 40% of MIP, and are drawn as reference points.

1. Hypercapnic patients had higher FRC/TLC than non-hypercapnic patients as shown in Fig. 4.

2. The MIP tended to be smaller in hypercapnic patients (Fig. 4).

3. There was a significant positive relationship between PaCO$_2$ and the pleural pressure swing (Pi/MIP) and lung resistance (RL/MIP) as shown in Fig. 5.

4. The prevalence of hypercapnic patients was a function of the RL/MIP ratio or Pi/MIP indexes.

This information provides a strong indication that a steady state chronic inspiratory load, particularly when requiring efforts approaching the fatigue threshold,[7] may elicit a negative feedback reflex by which further neuromuscular activity is withheld, breathing pattern results in higher VD/VT and hypercapnia develops. If this is so, chronic hypercapnia may be an index of imminent fatigue if further increases in ventilation are required, i.e. by increased exercise. Some physiotherapists for example, teach imposed diaphragmatic breathing patterns such as slow deep breathing, which may eventually lead to fatigue. The implication is that COPD with chronic hypercapnia could benefit more from periods of resting of the respiratory muscles via surface ventilator devices or via endotracheal ventilation in those who have a tracheostomy rather than impose exercise. Periodic

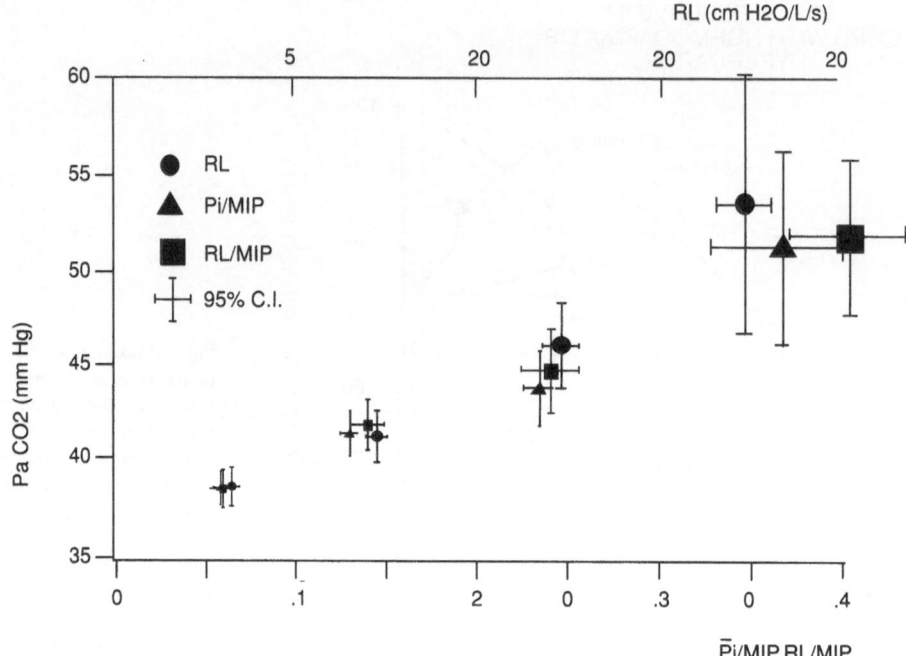

Fig. 5. Relationship between PaCo$_2$ breathing room air and lung resistance (RL), mean inspiratory pressure swing (Pi) (expressed as a fraction of maximal inspiratory pressure) (MIP) and the RL/MIP ratio. Data is average of 269 COPD patients.

mechanical ventilation may restore muscle function by their rest and contribute to the reduction of hypercapnia. If they become stronger, however, while resting the muscle, care should be taken not to hyperventilate excessively, causing a CO$_2$ wash-out.

Hence, chronic lung hyperinflation imposes an anatomical alteration to the chest wall by which inspiratory muscle force is decreased and respiratory loads could induce hypercapnia when the force required to breathe becomes near that required to fatigue the muscles. Lung hyperinflation is central to this mechanism by indirectly inducing inspiratory muscle dysfunction.

References

1. Robin E., O'Neill R.: The fighter versus the nonfighter. Arch. Environm. Health 1963; 7:125-129
2. Whitelaw W., Derenne J.P., Milic-Emili J.: Occlusion pressure as a measure of respiratory center output in conscious man. Resp. Physiol. 1975; 23:181-189
3. Sorli J., Grassino A., Lorange G.,Milic-Emili J.: Control of breathing in patients with COPD. Clin. Sci. Mol. Med. 1978; 54:295-304

4. Grassino A., Macklem P.T.: Respiratory muscle fatigue and ventilatory failure. Ann. Rev. Med. 1984; 35:626-647

5. Bellemare F., Grassino A.: Effect of pressure and timing of contraction on human diaphragmatic fatigue. J. Appl. Physiol. 1982; 53:1190-1195

6. Bellemare F.,Grassino A.: Evaluation of human diaphragmatic fatigue. J. Appl. Physiol. 1982; 53:1196-1206

7. Bellemare F., Grassino A.: Force reserve of the diaphragm in COPD patients. J. Appl. Physiol. 1982; 55:8-15

8. Skatrud J.B., Dempsey J.A., Bhansali P., Irvin C.: Determinants of chronic carbon dioxide retention and its correction in humans. J. Clin. Invest. 1980; 65:813-821

9. Pardy R.L., Roussos C.: Endurance of hyperventilation in COPD. Chest 1983; 83:744-750

10. Sadoul N., Bazzy R., Akabas S., G.G. Haddad: Ventilatory responses to fatiguing and non-fatiguing resistive loads in awake sheep. J. Appl. Physiol. 1985; 59:969-978

11. Laporta D., Grassino A.: Assessment of Pdi in humans. J. Appl. Physiol. 1985; 58:1469-1476

12. Martin J., Moreno R.H., Pare P., Pardy R.L.: Measurement of inspiratory muscle performance with incremental loads. Am. Rev. Respir. Dis. 1987; 135:919-923

13. Begin P., Grassino A.: Prevalence of CO_2 retention in COPD patients related to inspiratory muscle and lung dysfunction. Am. Rev. Respir. Dis. 1986;133: A191

13. Pattern of Systemic Venous Return during Negative Pressure Ventilation with Pneumowrap: A Pulsed Wave Doppler Study in Patients with COPD

A. TORBICKI[1], N. AMBROSINO[2], C. FRACCHIA[2], R. TRAMARIN[2], M. POZZOLI[2], C. RAMPULLA[1]

1. Department of Hypertension and Angiology, Medical Academy, Warsaw, Poland
2. "S. Maugeri Foundation" Pavia, Care and Research Institute, Medical Center of Rehabilitation, IRCCS, Fondazione Clinica del Lavoro, Montescano, Pavia, Italy

Introduction

Intermittent negative pressure ventilation (INPV) has been suggested as decreasing the fatigue of inspiratory muscles and the severity of respiratory failure in patients with obstructive pulmonary disease[2]. However, it is still not clear whether and to what extent the respiratory muscles are rested during INPV applied by means of cuirass and pneumowrap ventilators. The reliable data on the presence or absence of spontaneous contractions of the respiratory muscles would be important not only from the point of view of the effectiveness of muscular rest. By creating different changes in pleural and abdominal pressures INPV with and without active contribution of inspiratory muscles may differently affect both preload and afterload of the ventricles. Better understanding of the mechanical function of the diaphragm during INPV might be of help in explaining the contradictory results of studies addressing hemodynamic effects of this kind of respiratory support[3] (Ambrosino, unpublished information). EMGdi is not considered a fully reliable marker of the mechanical function of the diaphragm during supported ventilation. We attempted to identify active contractions of the diaphragm through their effect on the blood flow velocity in the inferior vena cava (IVC). We assumed that contracting the diaphragm should increase cranial flow velocity in IVC by a simultaneous increase in abdominal pressure and decrease in pleural pressure. This effect should be less marked, absent or inverted during INPV with a relaxed diaphragm, when negative pressure is externally applied both to the thorax and the abdomen (pneumowrap ventilators). We attempted to assess, with pulsed wave Doppler, the effect of inspirations on the pattern of venous flow velocity in the IVC prior to, during and directly after the INPV session by means of pneumowrap.

Materials and Methods

Studies were carried out in 6 patients with severe COPD and respiratory insufficiency and in one healthy volunteer (Table I).

Table I. Patients studied

Patient	Age	FEV$_1$/FVC	PaO$_2$ (mmHg)	PaCO$_2$ (mmHg)
F.E.	45	(*)	40	55
P.A.(**)	62	47%	48	48
B.G.	49	47%	48	64
A.R.	58	48%	43	58
D.M.	74	(*)	40	59
T.A.(**)	60	38%	46	50
C.R.(***)	46	-	-	-

*	the patient was not able to perform a reliable spirometric test
**	unsuccessful Doppler examination during negative pressure ventilation
***	healthy volunteer

All patients were in stable conditions at the time of the study, without exacerbation of respiratory symptoms. All patients had been receiving INPV treatment at least for 1 week (6 hrs/day, 5 days a week) before entering the study. Informed consent was obtained from all patients. The Doppler recordings were performed during the first hour of INPV by means of a pneumowrap. The ventilator was set to deliver negative pressures from -20 to -40 cm H$_2$O at a rate slightly higher than the patient's spontaneous breathing frequency. The echocardiographic transducer (Hewlett-Packard 2.5/1.9 MHz) together with the arm of the examiner were introduced into the pneumowrap using the leg approach. The transducer was activated and placed under two-dimensional echocardiographic control in the subcostal region so as to visualize the inferior vena cava (IVC) at its entrance to the right atrium (Fig. 1). Color-coded, two-dimensional echocardiography was used to indentify the mainstream of venous inflow to the right atrium and, finally, pulsed Doppler sample volume was placed within this mainstream.

Alternatively, the sample volume was placed within the upper hepatic vein using the same approach. Doppler tracings were recorded simultaneously with a noncalibrated chest plethysmogram, provided by a pulse transducer positioned under an elastic band wrapping the middle part of the thorax. The electrocardiographic electrodes were positioned in such a way as to record also electrical diaphragmatic activity during inspiration (Edi). The sound produced by the flow of air pumped out of the airtight suit was recorded with a phonocardiographic transducer and was used to identify the timing of mechanically produced inspiration (Fig. 2).

The study was performed before, during the first 60 minutes and directly after the INPV session. The transducer and the arm of the examiner remained in place during this whole period to avoid additional mechanical impulses in the diaphragmatic region. Efforts were made to provide psychosomatic relaxation of the patient.

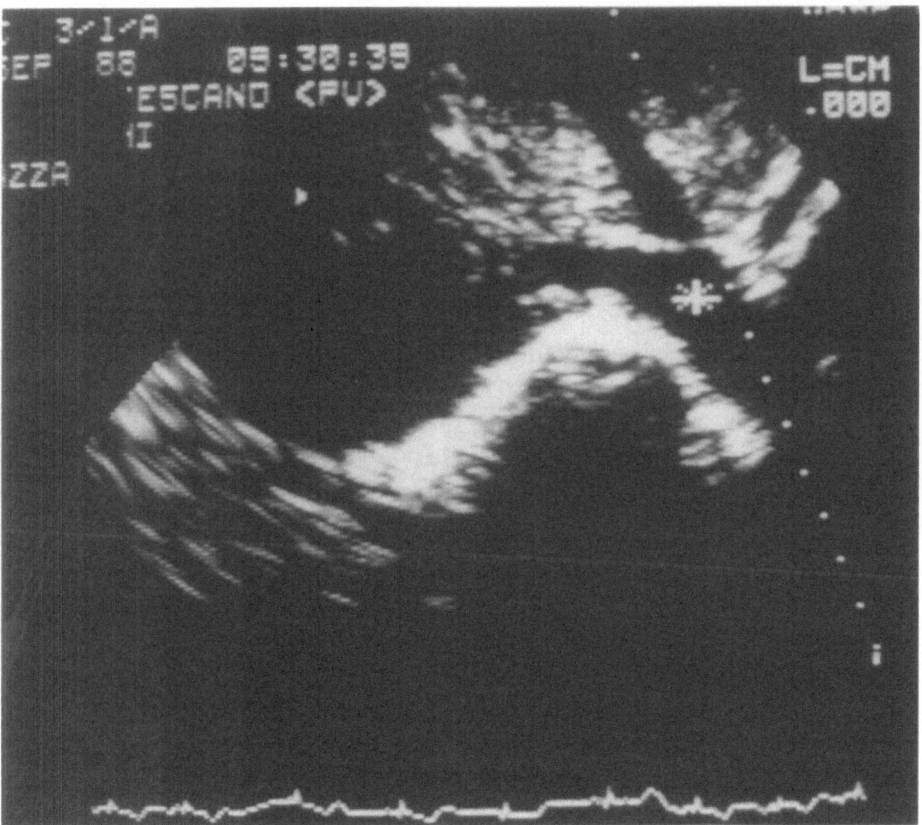

Fig. 1. Two-dimensional echocardiographic image of the region of confluence of hepatic veins at their entrance to the inferior vena cava (+). Apart from two-dimensional echocardiography, the position of pulsed wave Doppler sample volume in the mainstream of venous flow towards the right atrium was monitored with colour flow mapping.

104

Results

Registration of central venous flow during INPV was possible in 4 out of 6 patients. In 2 patients obesity and excessive height respectively did not permit us to place the transducer in the proper position inside the pneumowrap. During spontaneous breathing all patients showed a similar pattern of venous return. It was

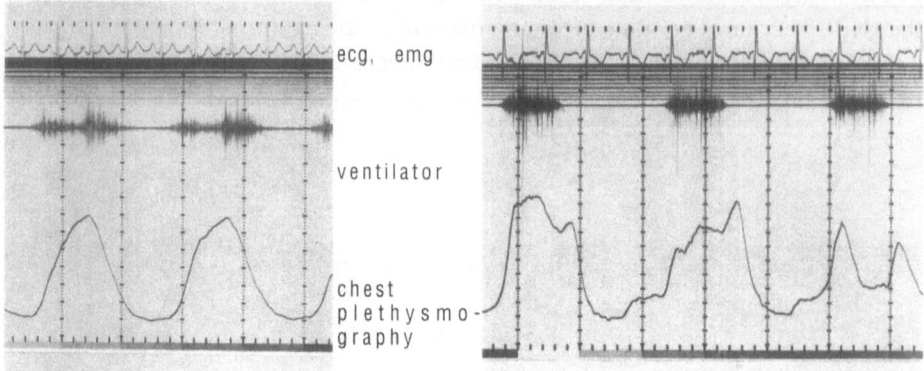

Fig. 2. Chest plethysmography, ECG/EMG and the function of the mechanical ventilator were used as reference for the timing of Doppler-detected venous flow velocity changes. Irregular plethysmographic curve in the right panel indicates interference of mechanical and spontaneous ventilation.

characterized by marked accelerations of cranial flow in the inferior vena cava and hepatic veins during each inspiration (Fig. 3).

The amplitude of respiration-related changes in venous flow velocity (VFV) clearly predominated over the amplitude of VFV waves dependent on the phases of the heart cycle. In a healthy volunteer, spontaneous respiration resulted in a pattern of VFV similar to the one observed in patients with COPD, though the ratio between the amplitude of the respiration-related and heart-cycle-related flow velocity changes was lower than in patients.

Despite a decrease in diameter of the IVC during inspiration the stroke volume of the right ventricle was still increased. This was evidenced by Doppler tracings recorded within the right ventricular inflow and outlow tract, showing increased integrals of flow velocity curves of the beats directly following each inspiration.

During INPV three main patterns of venous flow velocity could be identified. The first one (pattern A) was identical to the VFV pattern observed during spontaneous respiration. It occurred despite ventilatory effective INPV, as judged from the regular, high-amplitude sinusoid plethysmographic curve synchronous with the function of the ventilator. Termination of INPV had no effect on the venous flow pattern in these cases (Fig. 4).

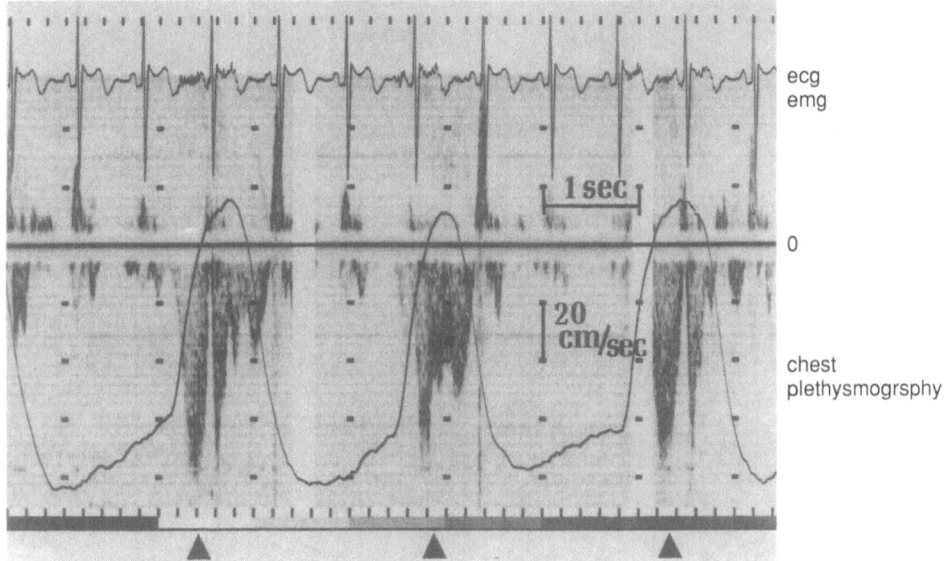

ecg
emg

0

chest
plethysmogrsphy

Fig. 3. Spontaneous respiration was characterized by marked inspiration-related accelerations of venous flow (arrows). Because the direction of blood which enters the right atrium from the inferior vena cava is away from the echocardiographic transducer, blood velocity is displayed below the zero line on the Doppler tracing.

Pattern A predominated during the first hour of INPV in 3 out of 4 patients. One of these three patients presented a modified VFV pattern A: isolated drop-outs of phasic venous flow accelerations were observed every 3-4 inspirations, despite effective mechanical ventilation (Fig. 5). Lower amplitude of plethysmographic and Edi waves during corresponding inspirations suggested momentary absence of contraction of inspiratory muscles.

A second VFV pattern (pattern B) coexisted with plethysmographic and electromyographic evidence of interference between spontaneous and mechanical ventilation. It was characterized by irregular increments of VFV, most of them occurring simultaneously with those chest expansions, which were clearly due to spontaneous inspirations (Fig. 6). This pattern was typical for the initial phase of INPV in all the patients and persisted throughout the whole period of INPV in one patient.

A modified pattern B was observed in 1 patient: despite high-amplitude, regular chest expansions due to the function of the ventilator, the VFV accelerations followed their own rhythm, which was slightly different from the frequency of mechanical ventilation (Fig. 7). Retrospectively, low-amplitude "artifacts" on the plethysmographic and electromyographic tracings were found accompanying each acceleration of venous flow, confirming the presence of the occult inspiratory muscular contractions.

In two patients, initially presenting type A and type B of VFV, respectively, the

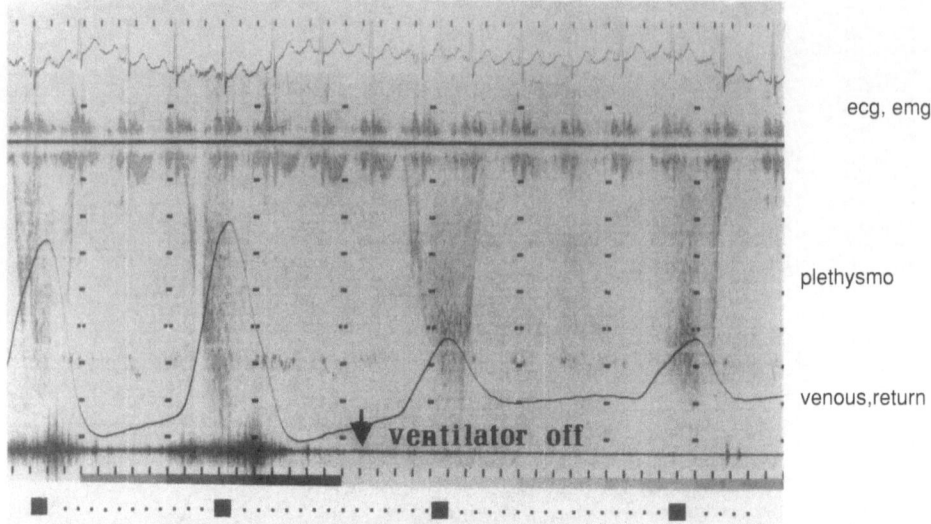

Fig. 4. Tracing of a patient with COPD (BG, 49 yrs): respiration-related venous flow velocity pattern not affected by the termination of mechanical ventilation (ventilator off). Blocks under the tracing follow the Doppler-detected pattern of venous flow velocity. Note lower frequency of breathing and lower amplitude of oscillations of the plethysmographic curve (plethysmo) after the termination of INPV.

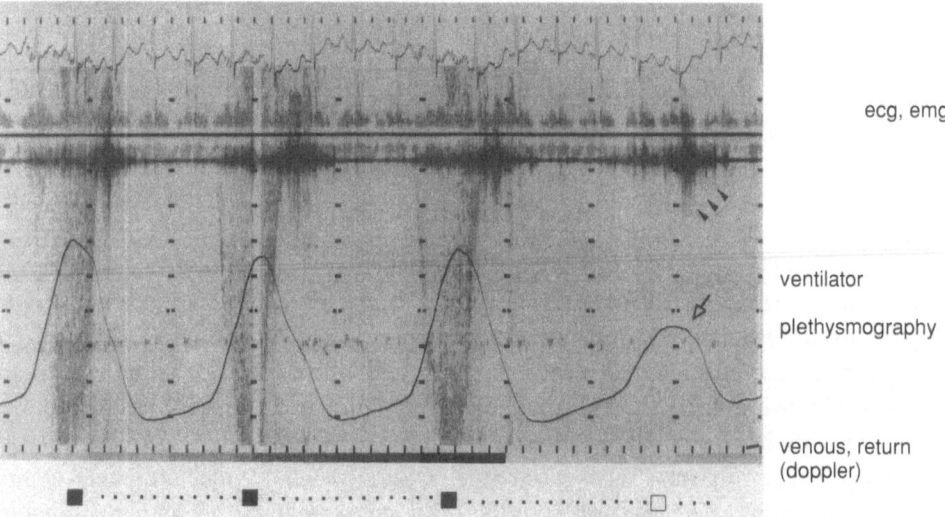

Fig. 5. The same patient as in Fig. 4: drop-out of inspiration-related acceleration of venous flow despite chest expansion during INPV (empty block below the tracing). Note the coinciding lower amplitude of the plethysmographic tracing (arrows) and lack of electromyographic "artifact" on the ECG tracing. Compare with Fig. 4.

Doppler examination during INPV was repeated after 6 months and 1 month respectively. In both patients VFV showed a similar pattern to the one observed during the first session.

Unlike the patients with COPD, respiration-related VFV changes disappeared completely starting from the fifth minute of effective INPV (pattern C) in a healthy volunteer (Fig. 8).

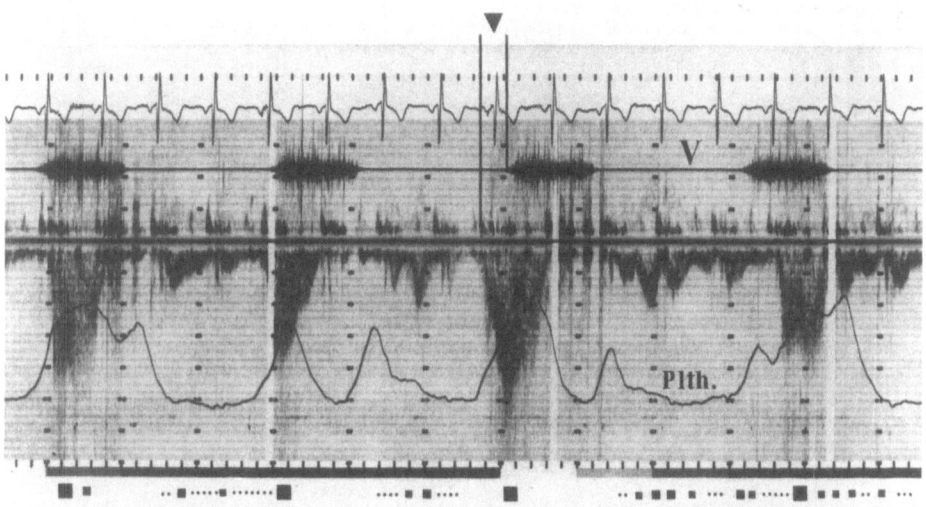

Fig. 6. Tracing of a patient with COPD (AR, 58 yrs): plethysmographic evidence of interference of mechanical (V) and spontaneous respiration. Acceleration of venous flow precedes the beginning of mechanical "inspiration" in the respiratory cycle indicated by the arrow.

Discussion

One of the controversies concerning the use of INPV in patients with COPD and muscular fatigue is whether inspiratory muscles are indeed rested during this kind of ventilatory support.[1,4]

Diminution of diaphragmatic electromyographic activity may not be an optimal marker of the mechanical function of the diaphragm, as it can be affected by modified topography of the thorax during mechanical ventilation. Our study suggests that the change in the abdominal-thoracic pressure gradient caused by the contraction of the diaphragm may be deduced from its influence on the VFV pattern. Though we did not directly measure esophageal and abdominal pressures, plethysmography together with Doppler tracings provided complementary information which seemed sufficient to identify active inspiratory muscle contractions.

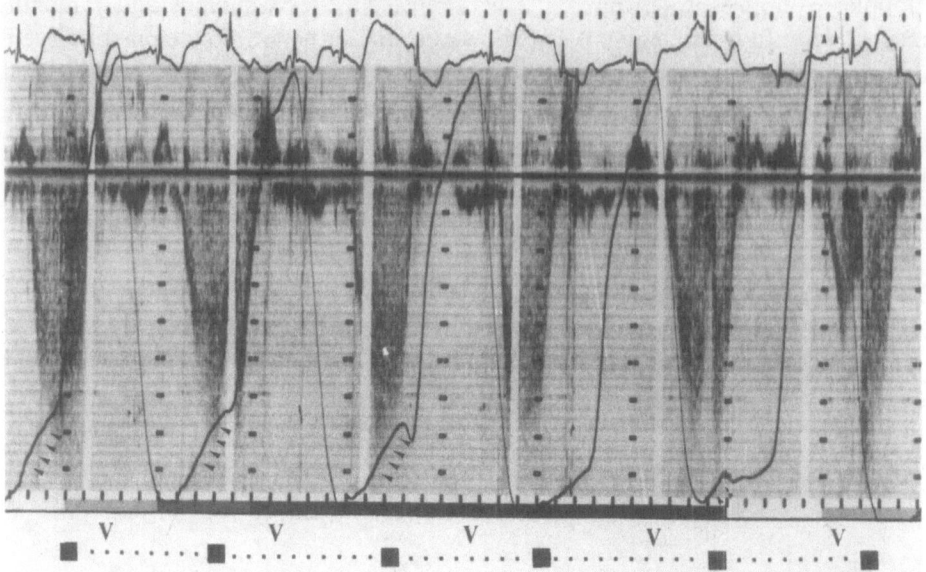

Fig. 7. Tracing of a patient with COPD (DV, 74 yrs): accelerations of venous flow (blocks) slightly asynchronous with the frequency of the high-amplitude oscillations of the chest plethysmographic curve caused by mechanical ventilation (V). Retrospectively, low-amplitude artifacts on the plethysmographic tracing (arrows) were found to coincide with venous flow accelerations.

Interestingly, Doppler and Edi were in agreement in identifying those inspirations, during which the diaphragm was relaxed. As both methods approached diaphragmatic function using different pathophysiological assumptions this concordance of results may indirectly confirm their validity as markers of diaphragmatic contraction.

In 1 out of 4 patients we found evidence of an unsynchronized interaction between mechanical and spontaneous respiration. In 3 patients the INPV was supported with active inspirations synchronized with the function of the ventilator. In only one patient was hemodynamic evidence of the abdominal-thoracic inspiratory pressure gradient repeatedly absent during isolated respiratory cycles. On the contrary, inspiration induced changes in VFV disappeared during INPV in a healthy volunteer, suggesting a lack of active diaphragmatic contractions.

Thus, our preliminary, qualitative observations suggest that the pattern of venous flow velocity assessed with pulsed wave Doppler may be useful in identifying active diaphragmatic contractions during INPV.

In contrast to a normal subject, patients with COPD seem to preserve this active contribution to inspiration during the first hour of INPV with the pneumowrap ventilator. However, certain drawbacks of our study must be recognized. The

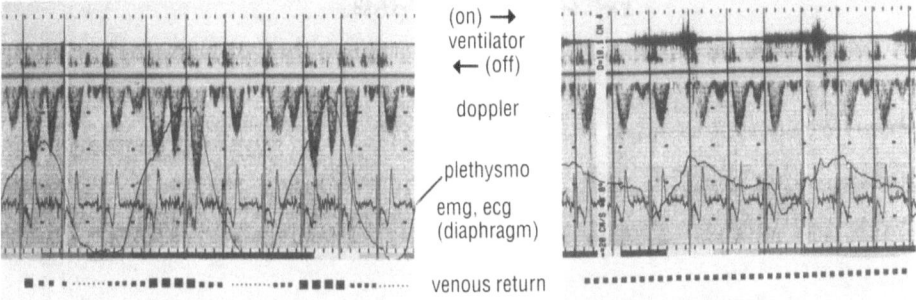

(on) →
ventilator
← (off)

doppler

plethysmo

emg, ecg
(diaphragm)

venous return

Fig. 8. Tracing of a healthy volunteer (CR, 46 yrs): respiration-related changes in venous flow velocity (represented by blocks below the tracings) are present only during spontaneous respiration (left panel) but not during negative pressure ventilation (right panel).

population observed was very small. The lack of simultaneous direct measurements of esophageal and abdominal pressure has been already discussed. Also, additional mechanical impulses produced by the echocardiographic transducer touching the skin in the epigastric region might have prevented complete relaxation of the diaphragm. Even so, the difference between the diaphragmatic reaction in patients and in the healthy volunteer, who suffered from the same mechanical disturbance, should be noted. Whether longer sessions, more negative pressure or different training in INPV would result in a complete drop-out of inspiratory muscular function also in patients with COPD remains to be investigated.

References

1. Cropp A., Di Marco A.F.: Effects of intermittent negative pressure ventilation on respiratory muscle function in patients with severe chronic obstructive pulmonary disease. Am. Rev. Respir. Dis. 1987; 135: 1056-1061

2. Macklem P.T.: Rest in the treatment of respiratory muscle fatigue. In: Grassino A., Fracchia C., Rampulla C., Zocchi L. (Eds.) *Respiratory muscles in chronic obstructive pulmonary disease.* Springer Verlag, London, 1987, pp 183-190

3. Murray R., Criner G., Becker P., Mendoza J., Rubin L.: Negative pressure ventilation impairs cardiac function in patients with severe COPD. Am. Rev. Respir. Dis. 136: A15 (abstr)

4. Rodenstein D.O., Stanescu D., Cuttitta M., Liistro G., Veriter C.: Ventilatory and diaphragmatic EMG responses to negative pressure ventilation in airflow obstruction. J. Appl. Physiol. 1988; 65: 1621-1626

14. Phosphorus Depletion in Limb and Respiratory Muscles of Patients with Chronic Obstructive Pulmonary Disease (COPD): A Preliminary Report

E. Fiaccadori,[1] E. Coffrini,[1] P. Vitali,[1] N. Ronda,[1] A. Guariglia,[1] C. Fracchia,[2] C. Rampulla,[2] N. Ambrosino,[2] T. Montagna,[2] L. Zocchi,[2] L. Bonandrini,[3] A. Borghetti[1]

1. Institute of Clinical Medicine and Nephrology, University of Parma, Italy
2. "S Maugeri Foundation", Pavia, Medical Center of Rehabilitation, IRCCS, Fondazione Clinica del Lavoro, Montescano, Pavia, Italy
3. Chair of Microsurgery, University of Pavia, Italy

Introduction

Phosphorus plays a key role in cellular biochemical reactions responsible for energy production, storage, and utilization and represents a major component of membranes and other cell structures: thus, the maintenance of normal phosphorus balance and both normal serum and cellular phosphorus levels is critical for the normal function of the organism.[1,2] In both experimental and clinical conditions hypophosphatemia and phosphorus depletion are associated with a wide spectrum of clinical syndromes such as myocardial failure, hepatocellular damage, hemolysis, platelet and leucocyte disfunction, osteomalacia and spontaneous fractures, hypoparathyroidism, impaired glucose tolerance, etc.[3] Moreover, neurologic and neuromuscular signs and symptoms such as ataxia, confusion, delirium, tremor, hyporeflexia, skeletal muscle weakness, and rhabdomyolysis have been observed.[1,3] Hypophosphatemia and/or phosphorus depletion have been equally indicated as possible determinants of respiratory muscle weakness in the course of respiratory failure.[4,5] In fact, an impairment of respiratory muscle contractile properties has been demonstrated in the course of hypophosphatemia, which improves with phosphorus repletion.[6,7] Hypophosphatemia is a common finding in the course of respiratory illnesses, with a prevalence of about 25%; in 5% of the same patients serum phosphate (Ps) levels may be extremely low (less than 1 mg% or 0.33 mmol/l).[8,9]

Recently, in a large series of 90 COPD patients, the prevalence of moderate (Ps < 2.5 mg or 0.83 mmol/l) and severe hypophosphatemia (Ps < 2.0 mg% or 0.66 mmol/l) was found to be respectively 14% and 9%;[10] moreover, in two patients with Ps < 2 mg% who underwent skeletal muscle needle biopsy, muscle phosphorus content (Pm) was reduced if compared to both normophosphatemic COPD patients and healthy control subject values.

No data are currently available concerning respiratory rauscle phosphorus content values in such conditions. The aims of this preliminary study were thus: (a) to evaluate muscle phosphorus content in respiratory muscles and in limb skeletal muscles in a group of patients with COPD and Chronic Respiratory Failure (CRF); (b) to determine the possible relationships between serum and skeletal muscle phosphorus in the same patients.

Patients and Methods

Thirteen patients (11M, 2 F, mean age 66 yrs ± 7 SD) with COPD and CRF ($PaCO_2$ 61 mmHg ± 8 SD; PaO_2 44 ± 7 SD) were studied. They were selected only if COPD was the primary diagnosis. COPD diagnosis had been made during previous hospitalizations or clinical visits and was based on positive history, clinical and radiological criteria, and standard measurements of pulmonary mechanics (FEV_1 values and FEV_1/FVC ratio less than 70% of predicted standards). No patient had impaired consciousness or was on parenteral nutrition. At the time of the study most patients were on active treatment with differently scheduled xanthine-derivatives, corticosteroids, diuretics and digitalis. No patient was diabetic or was receiving phosphorus supplements or antacids at the time of the study.

Serum creatinine levels ranged from 0.87 to 1.32 mg/dl. All patients underwent needle biopsies from the lateral portion of the quadriceps femoris muscle, according to the Bergstrom technique;[11] in five patients surgical biopsies from the external intercostal muscles (5th intercostal space, anterior axillary line) were also performed under local anesthesia.

Muscle samples (weighing 23-72 mg) were rapidly dissected free from visible fat or connective tissue and placed in preweighed quartz tubes; FFS (muscle fat-free solids) were then obtained, as previously described in detail.[12] Muscle fragments were digested with 0.200 ml of 70% $HClO_4$, and muscle total phosphorus content (Pm) was measured by the Chen method[13] and expressed as mmol/kg of FFS.

A group of twelve age-and sex-matched healthy subjects were utilized as controls for the quadriceps femoris study; external intercostal muscle surgical biopsies were obtained at anesthesia induction in another group of eleven subjects, with normal respiratory function, undergoing elective surgery. Serum phosphorus (Ps) was also measured in all patients and control subjects. All subjects and/or their next of kin were informed of the nature and the possible risks of the study, and consent was obtained from each participant and/or his or her relatives.

Statistics

Data are presented as mean ± SD. Student's "t" test for unpaired data was used to assess the significance of the differences between the means: standard techniques of linear regression and correlation by the least squares method were utilized. A statistical package (Statpak, Northwest Analytical Inc., Portland, OR) was used for calculations.

Results

Mean serum phosphorus levels of COPD patients were not significantly different from control values. Phosphorus content of both quadriceps femoris and external intercostal muscles was significantly reduced as compared to control values. (Fig. 1); no correlation was found between Pm values of the two kinds of muscle (Fig. 2). A significant (r=0.65, n=13, p<0.05) direct relationship was found between Ps and

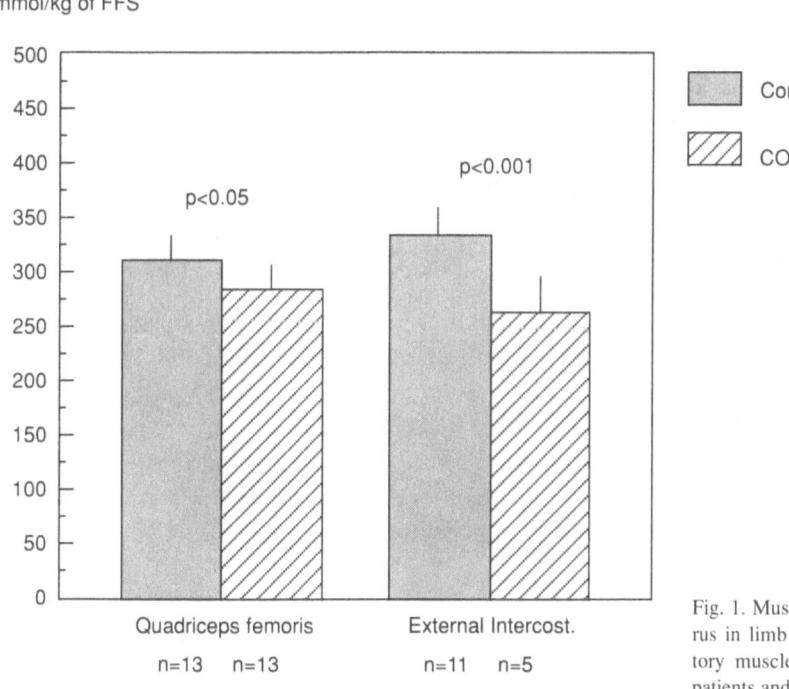

Fig. 1. Muscle phosphorus in limb and respiratory muscles of COPD patients and control subjects.

114

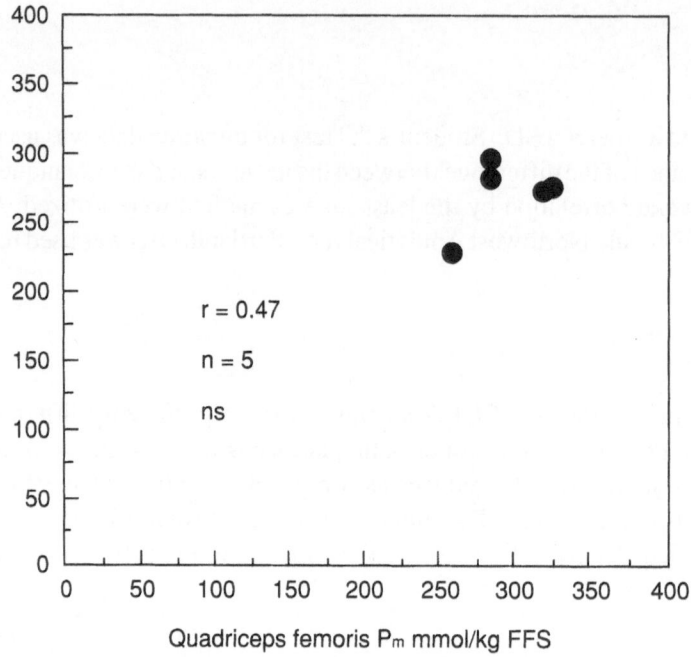

External Intercostals Pm mmol/kg FFS

r = 0.47

n = 5

ns

Quadriceps femoris Pm mmol/kg FFS

Fig. 2. Relationship between limb and respiratory muscle phosphorus (Pm) in COPD patients.

quadriceps femoris P_m (Fig. 3), but not between P_s and quadriceps femoris P_m (Fig. 3), and not between P_s and external intercostal muscles P_m (Fig. 4). No correlation was found between $PaCO_2$ or PaO_2 and P_m values of either quadriceps femoris or external intercostal muscles in COPD patients.

Discussion

Our preliminary report indicates that muscle phosphorus content may be reduced in both limb and respiratory muscles of patients with severe COPD. This abnormality probably reflects the influence of systemic factors negatively affecting skeletal muscle phosphorus metabolism, since it occurs to a similar extent in both resting limb and active respiratory muscles. Among these factors, nutritional depletion[14] and several drugs (xanthine derivatives, diuretics, corticosteroids, etc.) commonly used in course of COPD, all with a phosphaturic effect,[2,3,15,16] could play an important role.

In our patients, no correlation was found between serum phosphorus levels and respiratory muscle P_m, while hypophosphatemia was significantly related to low P_m content values in the quadriceps femoris muscle: however, in this latter case too, hyphophosphatemia seems to be a poor predictor of phosphorus depletion in COPD

Quadriceps femoris phosphorus mmol/kg FFS

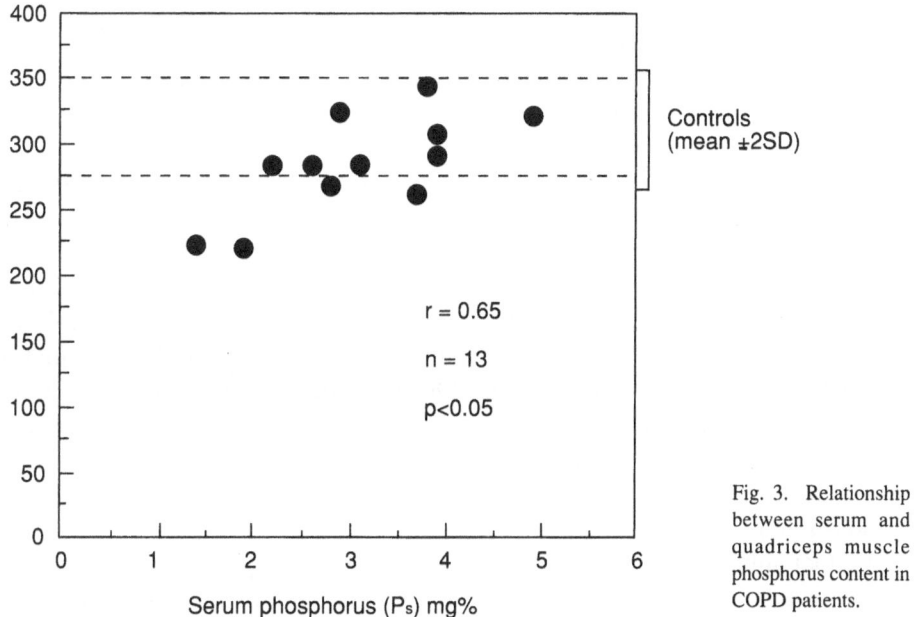

Fig. 3. Relationship between serum and quadriceps muscle phosphorus content in COPD patients.

External intercostal phosphorus mmol/kg FFS

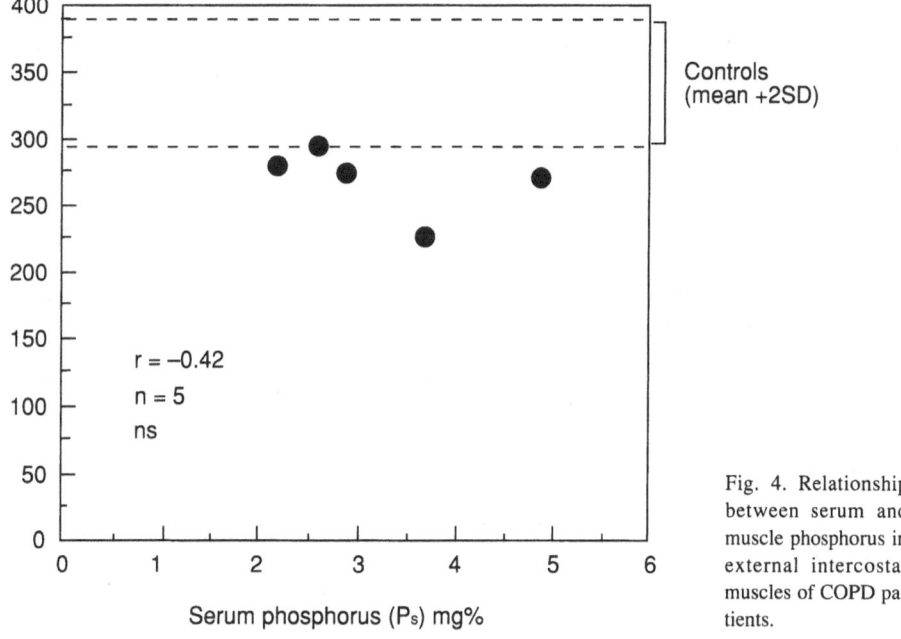

Fig. 4. Relationship between serum and muscle phosphorus in external intercostal muscles of COPD patients.

patients, since two patients of four having P_s values lower than 2.5 mg% showed P_m values in the normal range. Our results thus indicate that serum phosphorus levels are not predictive of muscle phosphorus content, although P_m may be reduced in COPD patients if severe degree hypophosphatemia is present.

Molecular mechanisms mediating the skeletal muscle myopathy of phosphorus depletion at cellular level have been recently discussed.[17] Abnormalities were found in the creatine phosphate energy shuttle required for cellular energy transport, in mitochondrial oxidative phosphorylation, and in myofibrillar energy utilization;[17] also, under experimental conditions, these events were preceded by a marked reduction in cellular phosphorus stores. Actually, in spite of the fact that markedly reduced levels of ATP and phosphocreatine have been observed in limb[18] and respiratory[19] muscles of COPD patients, the extent to which phosphorus depletion and/or hypophosphatemia are responsible for both muscle energy metabolism alterations and muscle functional impairment is not well defined. In COPD patients other factors such as intracellular acidosis,[12] magnesium depletion,[20] and malnutrition itself[14,21] may also contribute to the cell energy metabolism alterations demonstrated.

Further studies aimed at correlating both muscle phosphorus content and cell energy metabolism alterations with respiratory muscle function measurements are needed in order to assess the functional and clinical significance of phosphorus metabolism alterations in COPD.

References

1. Lee D.B.N., Kurokawa K.: Physiology of phosphorus metabolism. In: Maxwell M.H., Kleeman C.R., Narins R.G. (Eds.) *Clinical Disorders of Fluid and Electrolyte Metabolism*. New York, McGraw-Hill, 1987: 245-296

2. Brautbar N., Kleeman C.R.:Hypophosphatemia and hyperphosphatemia: clinical and physiological aspects. In: Maxwell M.H., Kleeman C.R., Naris R.G. (Eds.) *Clinical Disorders of Fluid and Electrolyte Metabolism*. New York, Mc.Graw-Hill, 1987: 789-830

3. Knochel J.P., Jacobson H.R.: Renal handling of phosphorus, clinical hypophosphatemia and phosphorus deficiency. In: Brenner B.R., Rector F.C. (Eds.) *The Kidney*. Philadelphia: W.B. Saunders Company, 1986:619-662

4. Grassino A.: Determinants of respiratory muscle failure. Am. Rev. Resp. Dis. 1986; 134: 1091-1093

5. Rochester D.F., Arora N.S. Respiratory muscle failure. Med. Clin. North Am. 1983; 67: 573-598

6. Aubier M., Murciano D., Lecocguic Y., Viires N., Jacquens Y., Squara P., Pariente R.: Effect of hypophosphatemia on diaphragmatic contractility in patients with acute respiratory failure. New Engl. J. Med. 1985; 313:420-424

7. Gravelyn T.P., Brophy N., Siegert C., Peters-Golden M.: Hypophosphatemia-associated respiratory muscle weakness in a general inpatient population. Am. J. Med.1988; 84: 870-876

8. Fisher J., Magid N., Kallman C., Fanucchi M., Klein L., McCarthy D., Roberts I., Schulman G.: Respiratory illness and hypophosphatemia. Chest 1983; 83: 504-508

9. Migliazzo A., Pigleton S.K.:Hypophosphatemia in a respiratory intensive care unit. Am. Rev. Resp. Dis. 1983; 131: 152A

10. Fiaccadori E., Coffrini E., Vitali P., et al.: Phosphorus depletion and hypophosphatemia in course of chronic obstructive pulmonary disease. It. J. Chest Dis. (In press)

11. Bergstrom J.:Muscle electrolytes in man. Scan. J. Clin. Lab. Invest. 1962; 14 (suppl. 68): 1-110

12. Fiaccadori E., Del Canale S., Arduini U., Antonucci C., Coffrini E., Vitali P., Melej R., Guariglia A.:Intercellular acid-base and electrolyte metabolism in skeletal muscle of patients with chronic obstructive lung disease and acute respiratory failure. Clin. Sci. 1986; 71: 703-712

13. Chen P.S., Toribara T.Y., Warner H.: Microdetermination of phosphorus. Analyt. Chem. 1956; 28: 1756-1758

14. Fiaccadori E., Del Canale S., Coffrini E., Vitali P., Antonucci C., Cacciani G.C., Mazzola I., Guariglia A.: Hypercapnic-hypoxemic chronic obstructive pulmonary disease (COPD): influence of severity of COPD on nutritional status. Am. J. Clin. Nutr. 1988; 48: 680-685

15. Colin A.A., Kraiem Z., Kahana L., Hochberg Z.: Effects of theophylline on urinary excretion of cyclic AMP, and phosphorus in normal subjects. Min. Electr. Metab. 1984; 10: 359-361

16. Fiaccadori E., Vitali P., Coffrini E., Guariglia A., Mazzola I., Ronda N., Cacciani G.C.,Borghetti A.: Effetti della teofillina sul metabolismo renale dei fosfati nei pazienti con Broncopneumopatia cronica ostruttiva. Medicina Toracica 1988; 10: 135-139

17. Brautbar N., Carpenter C., Baczynski R., Kohan R., Massry S.G.: Impaired skeletal muscle energetics in phosphate depleted rats. Kidney Int. 1983; 24: 53-57

18. Fiaccadori E., Del Canale S., Vitali P., Coffrini E., Ronda N., Guariglia A.: Skeletal muscle energetics, acid-base equilibrium and lactate metabolism in patients with severe hypercapnia and hypoxemia. Chest 1987; 92: 883-887

19. Gertz I., Hedenstierna G., Hellers G., Wahren J.: Muscle metabolism in patients with chronic obstructive lung disease and acute respiratory failure. Clin. Sci. Mol. Med. 1977; 52: 395-403

20. Fiaccadori E., Del Canale S., Coffrini E., Melej R., Vitali P., Guariglia A., Borghetti A.: Muscle and serum magnesium in pulmonary intensive care unit patients. Critical Care Medicine, 1988; 16(8): 751-760

21. Rochester D.F., Esau S.A.: Malnutrition and the respiratory system. Chest 1984; 85: 411-415

15. Mechanisms of Dyspnea

J.W. FITTING
Pneumology Department, Internal Medicine Department, Vaudois University Hospital, Lausanne, Switzerland

The word dyspnea, which refers to a respiratory sensation, comes from the Greek dus which means abnormal or difficult, and from pnein which means breathing. Thus, dyspnea could correspond to a sensation of either abnormal breathing or difficult breathing, the two meanings not being equivalent. The medical approach tends to differentiate the abnormal from the normal.

Thus, for the physician, dyspnea means a sensation of breathing which is abnormally hard for a given activity. The physician will usually not consider as dyspnea the sensation of breathing felt by a normal subject during exercise.

Conversely, the physiologist tries to identify the sensory mechanisms of dyspnea, which, until now, have not been shown to be different in normal subjects and in patients. Thus, in physiological studies it is customary to consider any sensation of difficult breathing as dyspnea, whether in a normal situation or not.

According to this definition, dyspnea can occur in very different conditions such as airway obstruction, pulmonary fibrosis, pulmonary vascular diseases, kyphoscoliosis, neuromuscular disorders, heart failure, anemia, or exercise and loaded breathing in normal subjects. One question which is unresolved is whether a single sensory mechanism or different mechanisms account for dyspnea in all these situations. As we shall see, most studies on respiratory sensations used loaded breathing, which enables us to determine the relationship between a quantified stimulus and the sensation it evokes.

Different origins have been considered for respiratory sensations such as dyspnea: the lungs, the upper airways, the rib cage, the respiratory muscles, and the chemical drive.

The lungs do not seem to play a role since the sensation evoked by an added load is not altered by a bilateral vagal block.[1] Besides, lung transplantation now offers a model of denervated lungs. Patients with heart-lung or double lung transplantation report experiencing dyspnea during hard exercise.

The upper airways may play a sensory role under certain circumstances. From several studies on load detection, it can be concluded that the upper airways are able to sense negative pressures, but that they are less sensitive than other structures in the chest, such as inspiratory muscles[2]. However, it is unlikely that the upper airways play a role in the dyspnea experienced by patients since external loading alone creates large subatmospheric pressures at this level. Afferents from rib cage receptors probably play a sensory role, since the perception of inspiratory loads is impaired in quadriplegic patients.[3]

The fact that respiratory muscles play a major role in sensations evoked by added respiratory loads has been established very clearly by two studies. In the first one, Killian et al.[4] compared the ability to detect very low resistances during spontaneous breathing and during passive ventilation by an external negative-pressure respirator. The detection threshold was markedly elevated during passive ventilation, indicating that load detection depends on active respiratory muscle contraction[4]. In the second study, Campbell et al.[5] studied the perception of a wider range of loads during normal breathing and during partial curarization. They found that the loads were consistently overstimated when the respiratory muscles were weakened by curare.[5]

If the respiratory muscles appear to play an important role in respiratory sensations evoked by added loads, and by inference in dyspnea, the question is: what are the relevant variables for these sensations?

In a given individual, dyspnea is closely linked to minute ventilation. For instance, if a subject goes from rest to graded exercise and back to rest, his dyspnea and his minute ventilation increase and decrease in close parallel. However, this close link can be altered in certain circumstances. If an asthmatic patient performs a given exercise once with a placebo and once with salbutamol, the ventilation profiles are similar in both circumstances but the dyspnea profiles are not. With salbutamol, dyspnea is less for any level of ventilation.[6]

Similarly, the link between dyspnea and minute ventilation breaks in the presence of an added external load. For any level of ventilation, dyspnea is stronger in the presence of an inspiratory resistance. It appears therefore that ventilation is not the determinant of sensation.

Killian et al.[4] determined which was the relevant variable for sensation in the face of different types of inspiratory loads. During breathing against resistive loads, the sensation increases not only with loads but also with inspiratory flow.

However, the sensation is uniquely related to inspiratory pressure. Similarly, during breathing against elastic loads, the sensation increases not only with loads but also with tidal volume. Again, the sensation is uniquely related to inspiratory pressure.[7] These experiments show that the perceived magnitude of added loads is not a function of the load itself, but is directly related to the pressure developed by inspiratory muscles.

The relationship between sensation and inspiratory pressure has been determined by using the techniques of psychophysics.

These techniques have allowed us to demonstrate that in all sensory modes the intensity of a sensation is related to the intensity of the stimulus by a power function: $\Psi = k\ \Phi^n$. This is known as Steven's law.[8] When the sensation increases linearly with the stimulus, as for the visual appreciation of dimensions, the exponent is equal to one. When the sensation increases more and more rapidly, as for electric shocks, the exponent is greater than one. And when the sensation increases less and less, as for the brightness of light, the exponent is lower than one.

For respiratory sensations, it was found that the perceived magnitude of loads (Ψ) is related to inspiratory pressure (P) and to its duration (T_x) according to the following formula: $\Psi = P1.3\ x\ T_x 0.56$. From the exponents, it appears that pressure is more important than the duration during which it is developed.[7]

If the inspiratory muscles play an important role in the sensations evoked by loads, the question of the receptors and neural pathways involved remains open. In limb muscles, both a sense of force and a sense of effort have been demonstrated. The sense of force is supposed to originate from muscle receptors, such as tendon organs or spindles. The sense of effort is supposed to originate from the efferent motor command, presumably through corollary discharges. However, the exact pathways are still unknown.[9] Normally, effort and force increase in parallel, but effort increases disproportionately to force when a muscle is weakened, as by operating at a suboptimal length, or during partial curarization or fatigue.

The distinction between effort and force applies to the respiratory muscles as well. Dyspnea may appear in three types of situations: increased ventilation as during exercise or pulmonary vascular disorders, increased respiratory impedance as in disorders with altered mechanics of the respiratory system, and muscle weakness as in neuromuscular disorders. Dyspnea could correspond to the sense of force of the respiratory muscles in the first two conditions but not in the last one, where respiratory pressures are low. In contrast, dyspnea could be explained by an increased sense of effort of the respiratory muscles in all three circumstances.[10]

Several observations support the hypothesis that dyspnea is related to the sense of effort of the respiratory muscles. For instance, large interindividual differences in dyspnea exist at similar levels of inspiratory pressure during resistive loading.

For the same pressure, dyspnea is higher for the subjects with weaker respiratory muscles who probably have to produce a greater effort.[11] We recently observed a patient with myasthenia gravis who received iterative plasmaphereses. Her dyspnea regularly increased as her respiratory muscles weakened, and vice versa. It is likely that her respiratory effort was higher when her respiratory muscles were weak, explaining her dyspnea.

The dyspnea experienced by patients with chronic hyperinflation, particularly with emphysema, is well explained by this sensory role of inspiratory muscles. Their respiratory impedance is increased because of high airway resistances. Their minute ventilation is often increased by increased dead space ventilation. The capacity of their inspiratory muscles is reduced as they operate at a shorter length and with an altered configuration because of hyperinflation. Furthermore, they are frequently malnourished. These factors may lead to fatigue which further weakens inspiratory muscles and thereby requires a higher respiratory effort.

Several questions remain to be answered. We do not know if the respiratory muscles are the sole contributors to dyspnea, and some observations suggest that they are not. For instance, it has been shown that dyspnea is higher during CO_2 rebreathing than during exercise, at any level of ventilation.[12] This observation poses the question of a sensory role for the chemoreceptors.

This question was addressed by Castele et al.[13] who hyperventilated normal subjects with a positive-pressure respirator with fixed values of tidal volume and frequency.

Then, the PCO_2 was progressively increased by raising the inspired fraction of CO_2, and the subjects had to signal when they first sensed a need for increased ventilation. They observed that the onset of inspiratory muscle activity, manifested by a fall in positive inflation pressure, always preceded the awareness of ventilatory need.[13] Thus, the possible sensation related to the chemoreceptors could not be separated from the sensation of muscle activity.

Finally, we do not know if the character of dyspnea is similar or variable in different respiratory diseases. A single sensory mechanism may be insufficient to explain different qualities of dyspnea. The sensation of dyspnea might stem from the motor command to respiratory muscles and be modulated by other factors as well.

References

1. Guz A., Noble M.I.M., Widdicombe J.G., Trenchard D., Mushin W.W., Makey A.R.: The role of vagal and glossopharyngeal afferent nerves in respiratory sensation, control of breathing and arterial pressure regulation in conscious man. Clin. Sci. 1966; 30: 161-170.
2. Gandevia S.C., Killian K.J., Campbell E.J.M.: The contribution of upper airway and inspiratory muscle mechanisms to the detection of pressure changes at the mouth in normal subjects. Clin. Sci. 1981; 60: 513-518.
3. Gottfried S.B., Leech I., DiMarco A.F., Zaccardelli W., Altose M.D.: Sensation of respiratory force following low cervical spinal cord transection. J. Appl. Physiol. 1984; 57: 989-994.

4. Killian KJ., Mahutte C.K., Campbell E.J.M.: Resistive load detection during passive ventilation. Clin. Sci. 1980; 59: 493-495.

5. Campbell E.J.M., Gandevia S.C., Killian K.J., Mahutte C.K., Rigg J.R.A.: Changes in the perception of inspiratory resistive loads during partial curarization. J. Physiol. London 1980; 309: 93-100.

6. Stark R.D., Gambles S.A., Chatterjee S.S.: An exercise test to assess clinical dyspnoea: estimation of reproducibility and sensitivity. Br. J. Dis. Chest 1982; 76: 269-278.

7. Killian K.J., Bucens D.D., Campbell E.J.M.: Effect of breathing patterns on the perceived magnitude of added loads to breathing. J. Appl. Physiol. 1982; 52: 578-584.

8. Steven S.S.: On the psychophysical law. Psychol. Rev. 1957; 64: 153-181.

9. McCloskey D.I.: Kinesthetic sensibility. Physiol. Rev. 1978; 58: 763-820.

10. Killian KJ, Campbell EJM. Dyspnea. In: C. Roussos, P.T. Macklem (Eds.) *The Thorax* Lung Biology in Health and Disease, New York, M. Dekker, 1985; 29: 787-828.

11. Jones G.L., Killian K.J., Summers E., Jones N.L.: Inspiratory muscle force and endurance in maximum resistive loading. J. Appl. Physiol. 1985; 58: 1608-1615.

12. Stark R.D., Gambles S.A., Lewis J.A.: Methods to assess breathlessness in healthy subjects: a critical evaluation and application to analyse the acute effects of diazepam and promethazine on breathlessness induced by exercise or by exposure to raised levels of carbon dioxide. Clin. Sci. 1981; 61: 429-439.

13. Castele R.J., Connors A.F., Altose M.D.: Effects of changes in CO_2 partial pressure on the sensation of respiratory drive. J. Appl. Physiol. 1985; 59: 1747-1751.

Therapeutic Approach

16. Pharmacological Management of Pulmonary Hyperinflation

R. Poggi, L. Dal Vecchio, M. Bernasconi, R. Brandolese, A. Rossi
Department of Anesthesia and Intensive Care, City Hospital, Institute of Occupational Medicine, University of Padua, Italy

Introduction

Whenever the time needed for a complete relaxed exhalation is shorter than the interval actually available between successive inspirations, the end-expiratory lung volume (EELV) remains above the elastic equilibrium volume of the respiratory system. This has been observed in animals[1] and in infants,[2] and in patients with chronic air flow obstruction (CAO) in stable conditions[3] and when they are acutely ill, whether mechanically ventilated[4,8] or spontaneously breathing.[9,10] When the respiratory muscles are relaxed, ventilation above the elastic equilibrium volume is unavoidably associated with "intrinsic" positive end-expiratory pressure (PEEPi),[5,11] and with a marked decrease in the operational length of the respiratory muscle which impairs their mechanical efficiency.[12]

In some conditions, e.g. infants and patients with pulmonary edema, a high EELV can improve gas exchange. By contrast, in CAO patients the functional residual capacity is increased because of the loss of lung elastic recoil and the further increase due to dynamic factors has harmful consequences in terms of ventilatory load and respiratory muscle contractility.[13] With acute exacerbation of the airway disease, in fact, pulmonary hyperinflation ensues and the discrepancy between the ventilatory load (work of breathing and PEEPi) and the mechanical efficiency of the respiratory muscles can lead to respiratory muscle fatigue and acute respiratory failure. We have found that in CAO patients with acute respiratory failure, during mechanical ventilation, the elastic equilibrium volume is well below the EELV, even more than 1 liter, and PEEPi may approach and exceed 20 cm H_2O.[7,8]

While the loss of lung elasticity constitutes irreversible anatomical damage, the dynamic component of pulmonary hyperinflation can be reduced; for example, the expiratory duration can be set by the ventilator as long as possible.

However, we have found that 15-20 s of relaxed expiration were barely sufficient to approach the elastic equilibrium, i.e. the expiratory flow becomes nil.[4,7,8] Clearly such an expiratory time is not compatible with adequate gas exchange and acid-base status. Therefore, decreasing airway resistance by systematic aspiration of secretions as well as using drugs active on bronchial smooth muscle is often the only valid mechanism to restore the conditions for spontaneous ventilation in mechanically ventilated CAO patients with acute respiratory failure (ARF).

Bronchodilators such as methylxanthines and β_2-adrenergic agonists are widely used in the therapy of acute bronchoconstriction[14-16] and the effects of bronchodilators on lung mechanics have been extensively studied in normal subjects[17,18] and in asthmatics.[19-21] By contrast, similar information is not available for mechanically ventilated patients with ARF, despite the fact that noninvasive methods to assess respiratory mechanics in mechanically ventilated patients have been recently developed[22] and well adapted to the operational and measuring devices of some modern ventilators.[7,23] Furthermore, there is a common conviction that bronchial obstruction is poorly reversible in CAO patients with ARF.

Patients and Methods

We studied nine and seven consecutive mechanically ventilated CAO patients before and after a bolus of 5-6 mg/kg to doxophylline i.v.,[24] and inhalation of fenoterol (0.4-0.8 to 1-2mg),[25] respectively.

Airway pressure (Paw), flow (V) and expired lung volume (VT) by electrical integration were measured with the pressure transducers of Servo 900C ventilator.[7,23] Brief end-expiratory and end-inspiratory airway occlusions were performed for direct measurement of PEEPi, as well as to compute respiratory compliance (Cst,rs) and resistance (Rrs,max and Rrs,min) according to previously described methods.[5,26]

Fig. 1 shows that, on average, both Rrs,max and Rrs,min decreased significantly following bronchodilators (-30% and -37% after doxophylline, and -28% and -33% after fenoterol, respectively). Cst,rs was essentially unchanged. PEEPi was significantly decreased, on average, after doxophylline (-41%) and after fenoterol (-56%). The arterial oxygen tension was slightly decreased by fenoterol aerosol from 105 ± 27.6 to 94.1 ± 15.3 mmHg, probably because of exaggeration of preexisting ventilation-perfusion mismatching.

These results not only show that both the bronchodilators used in these studies determined a marked decrease in respiratory resistance, thereby improving the rate of lung emptying and significantly reducing dynamic hyperinflation and intrinsic

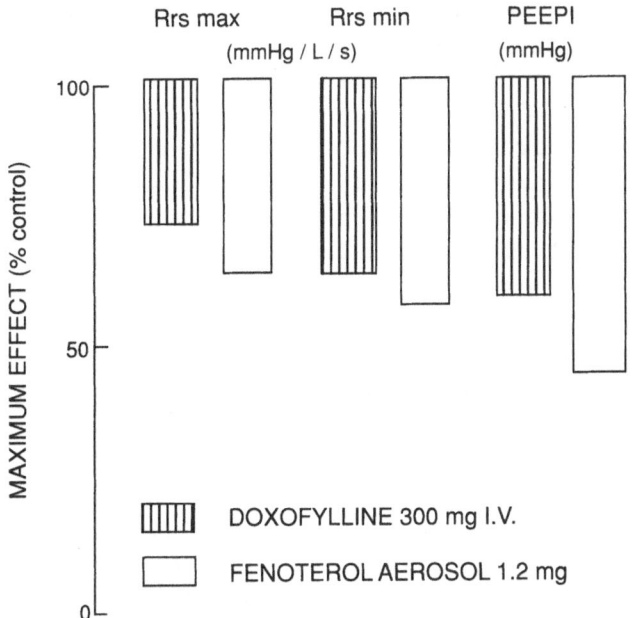

Fig. 1. Maximum effect (% of control) of doxophylline (dark columns) and fenoterol (white columns) on maximum and minimum respiratory resistance (Rrs,max and Rrs, min) and intrinsic PEEP (PEEPi), in nine and seven mechanically ventilated CAO patients.

PEEP, but also provide a useful insight into the mechanism of action. According to the analysis by Bates et al.,[27] in fact, changes in Rrs,min are mainly determined by changes in the airway caliber, whereas Rrs,max includes Rrs,min and the additional impedance of the periphery of the lung (i.e. the stress relaxation and the time constant inequalities). In our patients, the reduction in Rrs,max was mainly due to the decrease in Rrs,min (Fig. 1), indicating that also in critically ill CAO patients, and not only in asthmatics as is commonly accepted, there is a significant amount of reversible airway obstruction which can be obtained with drugs active on the airway smooth muscle.

Gay et al.[28] proposed that quick and simple assessment of the efficiency of bronchodilatation in mechanically ventilated patients can be obtained by measurements of changes in PEEPi and Pmax. However, Pmax is a rough measurement of respiratory function since it includes both the resistive and elastic components of airway pressure.[26] On average, also in our patients, changes in Pmax paralleled those of Rrs,max and Rrs,min, whereas Cst,rs essentially did not change. However, individual data may be more complex. In fact, in one patient there was a slight decrease in resistance (-5% for Rrs,max and -2% for Rrs, min), whereas Cst,rs was 31% higher following the doxophylline bolus; as a result, the 10% decrease in Pmax was determined by a decrease in the elastic pressure and not by a decrease in the resistive pressure. By contrast, in an other patient Cst,rs was 30% lower than control, while Rrs,max and Rrs,min decreased by 15% and 22% respectively; as a result, the decrease in resistance was offset by the contemporary decrease in compliance and Pmax was 4% higher 30 min after doxophylline.

The improvement in the mechanical properties of the total respiratory system has important implications in terms of patient-ventilator interaction and weaning.

In fact, as shown by Marini and coll.[29] the work of breathing done by the patient during assisted mechanical ventilation may not be negligible, and in extreme conditions, the patient can perform nearly as much work during a triggered cycle as during an unassisted spontaneous cycle. The decrease in resistance determined by bronchodilators decreased the resistive component of the work of breathing and therefore the mechanical load on the respiratory muscles when participating in inspiration. Moreover, the effort needed to receive the pressure boost during the pressure support ventilation[30] is also markedly reduced. In fact, although the negative "triggering" pressure is commonly set at few cmH_2O, this is on top of PEEPi, i.e. the end-expiratory elastic recoil, and the magnitude of the total negative pressure which has to be created by the respiratory muscles in order initiate the breath could be important and prevent adequate rest.[6] In our study bronchodilators markedly decreased PEEPi from 11.7 ± 5.4 to 6.9 ± 4.9 cmH_2O with doxophylline, and from 5.9 ± 3.9 to 2.6 ± 4.2 cmH_2O with fenoterol, on average. In this connection it should also be mentioned that methylxanthines can have a direct effect on respiratory muscle contractility, which should improve their mechanical output,[31] and therefore further support the respiratory muscle for weaning.

A point to be stressed in this analysis is that simple, non invasive, reproducible techniques are available for bed-side assessment of respiratory mechanics in mechanically ventilated patients, as well as to monitor the effects of therapy. These techniques are not time-consuming, can be well accepted by the critically ill patients, and are suitable for on-line display of physiological variables with computer equipped ventilators. Intrinsic PEEP and dynamic pulmonary hyperinflation have been well documented in CAO patients with acute respiratory failure. By contrast, few measurements can be found in stable CAO patients.[3] Fig. 2 shows a record from a representative CAO patient, during spontaneous tidal breathing. It can be seen that, as in CAO mechanically ventilated patients,[5] expiratory flow is present throughout expiration and is abruptly cut off at the end of expiration. The start of inspiratory flow is preceded by the onset of the negative pressure swing in Pes and Pdi records, indicating that the patient is dynamically hyperinflated and that in order to initiate inspiratory flow, the inspiratory muscles have to offset the elastic recoil pressure present at end expiration. Thus the "intrinsic" PEEP can be measured as the DPes between the onset of the pressure swing and the point corresponding to zero flow (Fig. 2). The constancy of Pga suggests that the diaphragm is acting more as a fixator and that other respiratory muscles (e.g. intercostals) are producing the actual inspiratory effort. In eighteen unselected CAO patients, PEEPi ranged from 0.8 to 6 cmH_2O. In ten patients, inhalation of 0.8

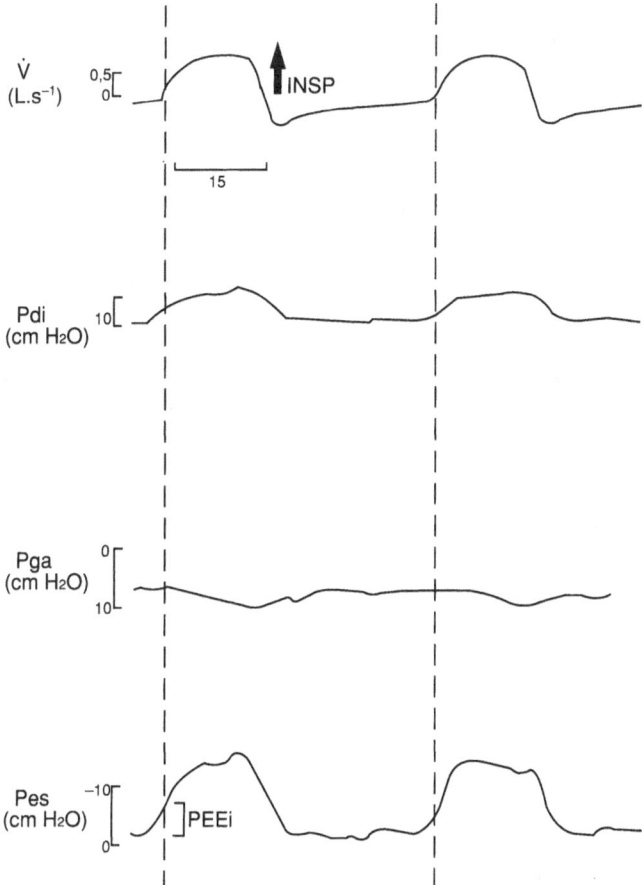

Fig. 2. From top to bottom: tracings of flow (V), transdiaphragmatic pressure (Pdi), gastric (Pga), and esophageal (Pes) pressure in a representative CAO patient during tidal breathing. Tidal volume is 0.75 l. Vertical dashed lines indicate the point corresponding to the onset of inspiratory flow. Note that in this patient expiratory flow abruptly ends before inspiration and Pes swings before the start of inspiratory flow. The difference between the onset of the negative swing and the point of zero flow represents the end-expiratory elastic recoil pressure, i.e. "intrinsic" PEEP (PEEPi), which has to be counterbalanced by the inspiratory muscles in order to start inspiration.

mg of fenoterol decreased PEEPi from 2.5 ± 1.5 to 0.9 ± 1.3 cm H_2O (-65.2% on average, $p < 0.01$), whereas in the placebo group (eight patients) PEEPi went from 2.4 ± 2.2 to 2.9 ± 2.6 cmH_2O (Fig. 3).

In the same patients we found that maximum transdiaphragmatic pressure (Pdi,max), a good index of diaphragmatic strength, significantly increased after fenoterol from 77.3 ± 16.4 to 91.1 ± 15.8 cmH_2O (mean +19%, range from -20% to +53%; $p < 0.05$).

With 0.2 mg of broxaterol, a recent β_2-adrenergic agonist, i.v., we found that a 35% increase in FEV_1 was associated with a 28% increase in Pdi,max.

The most likely explanation is that the marked bronchodilatation determined by adrenergic agonists in our CAO patients, improved the rate of lung emptying during resting breathing, virtually abolishing dynamic hyperinflation and lowered the end-expiratory lung volume, thereby lengthening the diaphragmatic fibers and increa-

132

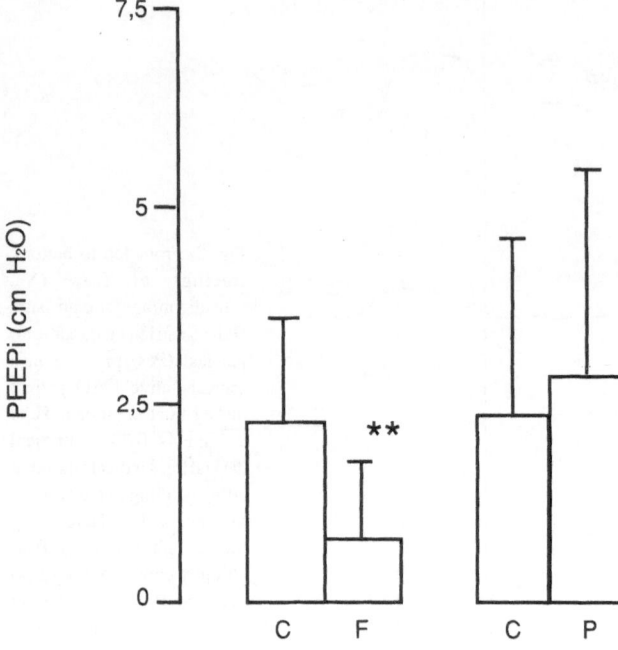

Fig. 3. Changes in intrinsic PEEP (PEEPi) after inhaled fenoterol (10 patients) and placebo (8 patients), in 18 stable CAO patients randomly assigned to the two groups. Columns are means and bars SD. C= control; F= fenoterol; P= placebo. ** p<0.01.

sing their mechanical output. However, we did not find any correlation between the intensity of bronchodilatation and the degree of improvement in maximum diaphragmatic strength. Therefore, on the basis of our data, we cannot exclude a direct action of adrenergic agonists on the diaphragmatic fibers.[32]

Conclusions

In summary we have shown that:

1) Some bronchodilators (e.g. doxophylline and fenoterol) can reverse airway obstruction at least in part, decreasing by the ventilatory load and dynamic hyperinflation in mechanically ventilated CAO patients with acute respiratory failure.

2) Changes induced by bronchodilatation in respiratory mechanics are important in improving the patient-ventilator interaction and in planning weaning, and can be used for simple, non-invasive bed-side monitoring.

3) The decrease in pulmonary dynamic hyperinflation may be associated with an improvement in respiratory muscle strength because they muscles can operate in a better part of their force-length relationship.

References

1. Vinegar A., Sinnett E.E., Leith D.E.: Dynamic mechanisms determine functional residual capacity in mice, musculus. J.Appl. Physiol. 1979; 46: 867-871.

2. Mortola J.:Dynamics of breathing in newborn mammals. Physiol. Rev. 1987; 67: 187-243.

3. Dal Vecchio L., Polese G., Broseghini C., Cestari E., Rossi A.: Effects of beta-2-adrenergic drugs of transdiaphragmatic pressure in patients with chronic obstructive pulmonary disease. Bull. Europ. Physiopath. Respir. 1987; 23(suppl 12):356.

4. Kimball W.R., Leith D.E., Robins A.G.: Dynamic hyperinflation and ventilator dependance in chronic obstructive pulmonary disease. Am. Rev. Respir. Dis. 1982; 126:991-995.

5. Rossi A., Gottfried S.B., Zocchi L., Higgs B.D., Lennox S., Calverley P.M.A., Begin P., Grassino A., Milic-Emili J.: Measurement of static compliance of total respiratory system in patients with acute respiratory failure during mechanical ventilation. Am. Rev. Respir. Dis. 1985; 131: 672-678.

6. Gottfried S.B., Rossi A., Milic-Emili J.: Dynamic hyperinflation, intrinsic PEEP, and the mechanically ventilated patient. Int. Critic. Care. Dig. 1986; 5: 30-33.

7. Broseghini C., Brandolese R., Poggi R., Polese G., Manzin E., Milic-Emili J., Rossi A.: Respiratory mechanics during the first day of mechanical ventilation in patients with pulmonary edema and chronic airway obstruction. Am. Rev. Respir. Dis. 1988; 138: 355-361.

8. Bernasconi M., Ploysongsang Y., Gottfried S.B., Milic- Emili J., Rossi A.: Respiratory compliance and resistance in mechanically ventilated patients with acute respiratory failure. Int. Care. Med. 1988; 14: 547-553.

9. Murciano D., Aubier M., Bussi S., Derenne J.P., Pariente R., Milic-Emili J.: Comparison of esophageal, tracheal, and mouth occlusion pressure in patients with chronic obstructive pulmonary disease during acute respiratory failure. Am. Rev. Respir. Dis. 1982; 126: 837-841.

10. Fleury B., Murciano D., Talamo C., Aubier M., Pariente R., Milic-Emili J.: Work of breathing in patients with chronic obstructive pulmonary disease in acute respiratory failure. Am. Rev Respir. Dis. 1985; 131-827.

11. Pepe P.E., Marini J.J.: Occult positive end-expiratory pressure in mechanically ventilated patients with airflow obstruction. Am. Rev. Respir. Dis. 1982; 126: 166-170.

12. Macklem P.T.: Hyperinflation. Am. Rev. Respir. Dis. 1984; 129:1-2.

13. Grassino A., Macklem P.T.: Respiratory muscles fatigue and ventilatory failure. Ann. Rev. Med. 1984; 35: 625-647.

14. Mitenko P.A., Ogilvie R.I.: Rational intravenous doses of theophylline. New Engl. J. Med. 1973; 289: 600-603.

15. Kelsen S.G., Kelsen D.P., Fleegler B.F., Jones R.C., Rodman T.: Emergency room assessment and treatment of patients with acute asthma. Am. J. Med. 1978; 64: 622-628.

16. Rossing TH, Fanta CH, Goldstein D.H., Snapper J.R., MacFadden E.R., Jr.: Emergency therapy of asthma: comparison of the acute effects of parenteral and inhaled sympathomimetics and infused aminophylline. Am. Rev. Respir. Dis. 1980; 122: 365-371.

17. Estenne M., Yernault J.C., De Troyer A.: Effects of parenteral aminophylline on lung mechanics in normal human. Am. Rev. Respir. Dis. 1980; 121: 967-971.

18. De Troyer A., Yernault J.C., Rodenstein D.: Influence of beta-2-agonist aerosols on pressure-volume characteristics of the lungs. Am. Rev. Respir. Dis. 1978; 118: 987-995.

19. Rossing T.H., Fanta C.H., McFadden E.R.: A controlled trial of the use of single versus combined-drug therapy in the treatment of acute episodes of asthma. Am. Rev. Respir. Dis. 1981; 123: 190-194.

20. Siegel D., Sheppard D., Gelb A., Wemberg P.F.: Aminophylline increases the toxicity but not the efficacy of an inhaled beta-adrenergic agonist in the treatment of acute exacerbations of asthma. Am. Rev. Respir. Dis. 1985; 132: 283-286.

134

21. Fanta C.H., Rossing T.H., McFadden E.R, Jr.: Emergency room treatment of asthma. Am. J. Med. 1982; 72: 416-422.

22. Milic-Emili J., Gottfried S.B., Rossi A.: Non-invasive measurement of respiratory mechanics in ICU patients. Int. J. Clin. Monit. 1987; 4: 11-20.

23. Jonson B., Nordstrom L., Olsson S.G., Akerback D.: Monitoring of ventilation and lung mechanics during automatic ventilation. A new device. Bull. Physiopath. Resp. 1975; 11: 729-743.

24. Poggi R., Bernasconi M., Brandolese R., Manzin E., Rossi A.: Effects of methylxanthines on respiratory mechanics in mechanically ventilated patients with chronic airway obstruction. Int. Care Med. 1988;14 (suppl 1):273.

25. Bernasconi M., Brandolese R., Massara A., Manzin E., Rossi A.: Effects of beta-2-adrenergic aerosol on respiratory mechanics in critically ill patients. Int. Care Med. 1988; 14 (suppl 1): 284.

26. Rossi A., Gottfried S.B., Higgs B.D., Zocchi L., Grassino A., Milic-Emili J.: Respiratory mechanics in mechanically ventilated patients with respiratory failure. J. Appl. Physiol. 1985; 58: 1849-1858.

27. Bates J.H.T., Rossi A., Milic-Emili J.: Analysis of the behaviour of the respiratory system with constant inspiratory flow. J. Appl. Physiol. 1985; 58: 1840-1848.

28. Gay P.C., Rodarte J.R., Tayyab M., Hubmayr R.D.: Evaluation of bronchodilator responsiveness in mechanical ventilated patients. Am. Rev. Respir. Dis. 1987; 136: 880-885.

29. Marini J.J., Capps J.S., Culver B.H.: The inspiratory work of breathing during assisted mechanical ventilation. Chest 1985; 87: 612-618.

30. Brochard L., Pluskwa F., Lemaire F.: Improved efficacy of spontaneous breathing with inspiratory pressure support. Am. Rev. Respir. Dis. 1987; 136: 411-415.

31. Aubier M., De Troyer A., Sampson M., Macklem P.T., Roussos C.: Aminophylline improves diaphragmatic contractility. New Engl. J. Med 1981; 305: 249-252.

32. Aubier M., Roussos C.: Pharmacotherapy. In: Roussos C., Macklem PT. (Eds.) The Thorax New York, M. Dekker, 1985; 29:1373-1405.

17. Mechanical Ventilation and Hyperinflation

J. MILIC-EMILI[1], A. ROSSI[2]

1. Department of Physiology, Meakins-Christie Laboratories, McGill University, Montreal, Quebec, Canada
2. Department of Anesthesia and Intensive Care, City Hospital, Institute of Occupational Medicine, University of Padua, Italy

Introduction

When normal individuals breathe at rest, the end-expiratory lung volume corresponds to the relaxation volume (Vr) of the respiratory system (also called elastic equilibrium volume because it represents the volume at which the elastic recoil pressure of the total respiratory system - Pel,rs - is zero). In this case, the inspiratory muscles must overcome two mechanical loads during lung inflation, namely the elastance and the flow resistance of the total respiratory system. By contrast, in patients with severe chronic airway obstruction (CAO) the resting end-expiratory lung volume usually exceeds Vr, a condition referred to as *dynamic hyperinflation*.[1] In this case, the end-expiratory lung volume does not correspond to the elastic equilibrium volume (Pel,rs=0) but a positive Pel,rs is present. This has been termed auto PEEP[2] or intrinsic PEEP (PEEPi).[3]

Intrinsic PEEP acts as an additional inspiratory load, i.e., as a *threshold* load to the inspiratory muscles.[4]

The magnitude of PEEPi depends on the time available for expiration (Te) and on the forces governing expiration.[5] At rest, expiratory muscle activity is usually absent, and hence the pressure driving expiration is due entirely to the elastic recoil pressure stored in the lung and chest wall during the preceding inspiration. This pressure is used in part to overcome the flow resistance offered by the total respiratory system during expiration (Rrs) while part of it is normally dissipated against the post-inspiration activity of the inspiratory muscles. Although normal subjects exhibit substantial braking due to activity of inspiratory muscles early in expiration, they are still capable of reaching Vr before the onset of the next inspiration because Rrs is normally relatively low.[5] By contrast, in CAO patients the rate of lung emptying is unduly slowed by the increased airway resistance and, by necessity, they develop PEEPi. When CAO patients become acutely ill (acute respiratory failure), they invariably exhibit tachypnea which implies a shortening of Te.[4] This further promotes dynamic hyperinflation.[1] Dynamic pulmonary hyperinflation not only implies increased inspiratory loading but also results in a reduction of the effectiveness of the inspiratory muscles as pressure generators.[6]

In this connection, it should be noted that the increase in FRC in CAO patients reflects in part dynamic hyperinflation and in part an increase in Vr due to loss of pulmonary elastic recoil (emphysema). Such an increase in Vr may be termed *static pulmonary hyperinflation*. The combination of an increased load to breathe and decreased pressure generating capacity of the inspiratory muscles may lead to respiratory failure due to diaphragmatic fatigue[7] which necessitates mechanical ventilation or other interventions designed to reduce the activity of the respiratory muscles. Providing rest for the respiratory muscles, thus allowing recovery from fatigue, is one of the putative benefits of mechanical ventilation. However, recent studies indicate that even during assisted mechanical ventilation the patient may still be making unduly large inspiratory efforts, particularly in the presence of high PEEPi values.[8,9]

Indeed, values of PEEPi as high as 22 cm H_2O have been reported in mechanically ventilated CAO patients admitted to the ICU for acute respiratory failure.[10]

This problem can be counteracted by suitable application of PEEP during assisted mechanical ventilation. In fact, in some instances it has also been possible to reduce substantially the breathing efforts in acutely ill CAO patients by appropriate application of CPAP or continuous negative pressure around the thorax (CNTP) without the need for intubation and mechanical ventilation.[11]

Strictly speaking, this approach is valid only in patients who exhibit dynamic expiratory flow limitation such that the application of CPAP or CNTP does not result in a further increase in the end-expiratory lung volume.

Indeed, if expiration is limited by dynamic compression of the airways, a decrease in transthoracic driving pressure due to an increase in airway pressure (CPAP) or a decrease in pressure around the thorax (CNTP) should not affect the rate of lung emptying[1] and, as a result, the end-expiratory lung volume should not change.[11]

If application of CPAP or CNTP does not change the end-expiratory lung volume, it is axiomatic that the inspiratory muscle pressure required to initiate lung inflation should be reduced by an amount approximately equal to PEEPi-CPAP or PEEPi+CNTP, where PEEPi is the end-expiratory elastic recoil pressure before application of CPAP or CNTP. Thus, appropriately dosed, CPAP, CNTP or PEEP should reduce the inspiratory threshold load due to PEEPi without affecting the FRC and hence the effectiveness of the inspiratory muscles as pressure generators.

Therapeutic Use of PEEP and CPAP

Since the original description in 1967 of the Adult Respiratory Distress Syndrome (ARDS),[13] mechanical ventilation with a positive end-expiratory pressure (PEEP) has become one of the most widely used and accepted techniques to improve pulmonary oxygenation in ARDS patients.[14]

It is now well established that PEEP and CPAP have neither a direct therapeutic action on the acutely injured lung[15] nor a preventive effect in patients at risk to of developing ARDS.[16]

In some ARDS patients, PEEP decreases the respiratory compliance[17] and increases the FRC[18] by recruiting previously collapsed, unventilated, perfused airspaces.[19,20] In recent years there has been tendency to apply lower levels of PEEP in ARDS patients than in the past, i.e., the "best" PEEP is set as "minimal" PEEP,[15,21] because high levels of PEEP are known to decrease the cardiac output and hence the O_2 delivery to the tissues of the body.

In the past, application of PEEP to CAO patients was not recommended in order to avoid barotrauma due to enhancement of pulmonary hyperinflation.[14] As discussed above, however, it is clear that appropriate administration of PEEP or CPAP may be beneficial to CAO patients by providing support for the inspiratory muscles in the patient-ventilator interaction[11] or during weaning.[22]

The interrelationship between PEEP, CPAP and CNTP with PEEPi is complex and not fully understood. However, recent studies support the notion that application of PEEP at levels lower than the initial PEEPi on ZEEP (zero end-expiratory pressure) reduces the work of breathing,[9,11] without significant adverse effects on thoraco-abdominal dimensions and configuration,[11,23] respiratory mechanics and cardiac output.[24]

In fact, in the presence of dynamic expiratory flow limitation, a well recognized corollary of advanced CAP,[25,26] the "waterfall" phenomenon, occurs such that an increase in downstream impedance (e.g. the application of external PEEP) relative to the site of flow limitation should have little effect on expiratory flow until the applied PEEP exceeds the initial level of PEEPi.[12]

Thus the FRC does not change, and hence there is no adverse effect also on the pressure generating capacity of the inspiratory muscles.

In conclusion, the application of CPAP or CNTP in patients with severe CAO appears to provide a useful strategy for reducing the inspiratory muscle efforts.

Application of PEEP should also be useful in CAO patients during assisted mechanical ventilation. It should be stressed, however, that it is imperative to monitor the lung volume when using these therapeutic procedures. This can easily be done by use of inductive plethysmographs ("resistance").

138

References

1. Gottfried S.B., Milic-Emili J.: Dynamic hyperinflation, intrinsic PEEP, and the mechanically ventilated patient. Int. Crit. Care Digest 1986; 5: 30-33
2. Pepe P.E, Marini J.J.: Occult positive end-expiratory pressure in mechanically ventilated patients with airflow obstruction. Am. Rev. Respir. Dis.1982; 126: 166-170
3. Rossi A., Gottfried S.B., Zocchi L., Higgs B.D., Lennox S., Calverley P.M.A., Begin P., Grassino A., Milic-Emili J.: Measurement of static compliance of total respiratory system in patients with acute respiratory failure during mechanical ventilation. Am. Rev. Respir. Dis. 1985; 131: 672-678
4. Fleury B., Murciano D., Talamo C., Aubier M., Pariente R., Milic-Emili, J.: Work of breathing in patients with chronic obstructive pulmonary disease in acute respiratory failure. Am. Rev. Respir. Dis. 1985; 132: 822-827
5. Shee C.D., Ploysongsang Y., Milic-Emili, J.: Decay of inspiratory muscle pressure during expiration in conscious humans. J. Appl. Physiol. 1985; 58: 1859-1865
6. Pengelly L.D., Alderson A.M., Milic-Emili J.: Mechanics of the diaphragm. J. Appl. Physiol. 1971; 30: 797-805
7. Bellemare F., Grassino A.: Effect of pressure and timing of contraction on human diaphragm fatigue. J. Appl. Physiol. 1982; 53: 1190-1195
8. Marini J.J., Rodriguez R.M., Lamb V.: The inspiratory work load of patient-initiated mechanical ventilation. Am. Rev. Respir. Dis. 1986; 134: 902-906
9. Smith T.C., Marini J.J.: Impact of PEEP on lung mechanics and work of breathing in severe airflow obstruction. J. Appl. Physiol. 1988; 65: 1488-1499
10. Broseghini C., Brandolese R., Poggi R., Polese G., Manzin E., Milic-Emili J., Rossi A.: Respiratory mechanics during the first day of mechanical ventilation in patients with pulmonary edema and chronic airway obstruction. Am. Rev. Respir. Dis. 1988;138: 355-361
11. Smikovitz P., Brown K., Goldberg P., Milic-Emili J., Gottfried S.B.: Interaction between intrinsic and externally applied PEEP during mechanical ventilation. Am. Rev. Respir. Dis. 1987;135: A202
12. Hyatt R.E.: Expiratory flows limitation. J. Appl. Physiol. 1983; 55: 1-8
13. Ashbaugh D.G., Bigelow D.B., Petty T.L., Levine B.E.: Acute respiratory distress syndrome in adults. Lancet 1967; 2: 319-323
14. Ashbaugh D.G., Petty T.L.: Positive end-expiratory pressure. J. Thor. Cardiovas. Surg.1973; 65: 165-170
15. Albert R.K. :Least PEEP: Primum non nocere. Chest 1985; 87: 2-3
16. Pepe P.E., Hudson L.D., Carrico C.J.: Early application of positive end-expiratory pressure in patients at risk for the adult respiratory distress syndrome. New Eng. J. Med. 1984; 311: 281-286
17. Suter P.M., Fairley H.B., Isenberg M.D.: Optimum end-expiratory airway pressure in patients with acute pulmonary failure. New. Eng. J. Med. 1975; 292: 284-289
18. Falke K.J., Pontoppidan H., Kumar A., Leith D.E., Geffin B., Laver, M.B.: Ventilation with end-expiratory pressure in acute lung disease. J. Clin. Invest. 1972; 51: 2315-2323
19. Matamis D., Lemaire F., Hrf A., Brun-Buisson, C., Ansquer, J.C., Atlan, G.: Total respiratory pressure-volume curves in the adult respiratory distress syndrome. Chest 1984; 86: 58-66
20. Gattinoni L., Pesenti A., Avalli L., Rossi F., Bombino M.: Pressure-volume curve of total respiratory system in acute respiratory failure. Conputed Tomographic Scan Study. Am. Rev. Respir. Dis. 1987;136: 730-36
21. Carrol G.C., Tuman K.J., Braverman B., Logas W.G., Woll N., Goldin M., Ivankovich A.D.: Minimal positive end-expiratory pressure (PEEP) may be "Best PEEP". Chest 1988; 93: 1020-1025
22. Milic-Emili J.: Is weaning an art or a science? Am. Rev. Respir. Dis. 1986; 134: 991-995,
23. Gottfried S.B., Simkovitz P., Skarbuskis M.: Effect of constant negative extrathoracic pressure

on breathing pattern and respiratory muscle function in severe chronic obstructive pulmonary disease (COPD). Am. Rev. Respir. Dis. 1987;135: A201

24. Brandolese R., Bernasconi M., Poggi R., Broseghini C., Manzin E., Milic-Emili J., Rossi A.: Effects of PEEP on respiratory mechanics and gas exchange in mechanically ventilated patients with ARDS and acute exacerbation of COPD. Am. Rev. Respir. Dis. 1988;137: A470

25. Gottfried S.B., Rossi A., Higgs B.D., Calverley P.M.A., Zocchi L., Bozic C., Milic-Emili, J.: Non-invasive determination of respiratory system mechanics during mechanical ventilation for acute respiratory failure. Am. Rev. Respir. Dis.1985; 131: 414-420

26. Kimball W.R., Leith D.E., Robins A.G.: Dynamic hyperinflation and ventilator dependence in chronic obstructive pulmonary disease. Am. Rev. Respir. Dis. 1982;126: 991-995

18. Chest Physiotherapy and Hyperinflation

R. SERGYSELS, A. LACHMAN, G. CAMINITI, R. WILLEPUT
Pneumology Department, Saint-Pierre University Hospital, Brussels, Belgium

Introduction

Hyperinflation is a consequence of mechanical adaptation to changes in pressure volume relationships of the chest wall and/or the lung parenchyma. In some circumstances, hyperinflation may also result from persistent tonic activity of the inspiratory muscles during the whole respiratory cycle.

Hyperinflation results in decreased resistive work but increased elastic work for breathing while inspiratory muscles are in an unfavourable length-tension relationship.[1-2] This leads to a breathing pattern which is rapid and shallow and therefore the VD/VT ratio is largely increased.

From a theoretical point of view, if hyperinflation "has to be corrected" by chest physiotherapy, then four techniques may be of interest.

Low Frequency Breathing with Active Expiration

Low frequency breathing has been exhaustively studied[3] but in order to lower the end expiratory volume the patients have to prolong their expiration by using their expiratory muscles.

In our experience[4-5] this results in increasing VT, lowering frequency without significantly improving gas exchange (PaO_2, $PaCO_2$). VT/TI and Ti/TOT are not significantly affected and VO_2 is slightly increased (NS). Magnetometric studies[6] show, in such circumstances, increased abdominal participation and in a number of patients, paradoxical motion of the chest wall (inwards inspiratory motion of the thorax, 2 out of 11 patients studied, outwards expiratory motion of the thorax, 9 out of 11 patients).

When abdominodiaphragmatic respiration was performed with fast relaxation of the abdomen at the beginning of inspiration, then we observed[7] in all subjects (normals and COPD) an inspiratory motion of the thorax but also a negative deflection of Ppl and Pgas allowing us to calculate during the beginning of inspiration, but only in normals, no change in Pdi from end expiratory volume until FRC.

This suggests that this type of breathing not only improves the curvature of the diaphragm at end expiration but also that the initiation of inspiration may be facilitated by relaxation of the abdomen.

However, the chest physiotherapist has to remember the recent work of Bellemare and Grassino,[8] showing that changing the breathing pattern and especially increasing TI/TOT may induce fatigue in patients with severe airflow obstuction.

Body Positioning

The only body position that may significantly decrease FRC in normals as in COPD patients is the supine position. Few patients describe a decrease of breathlessness in this position but Dretz and Sharp[9] described "better efficiency of the diaphragm" in that position. Patients often adapt a sitting forward position and again Dretz and Sharp[9] could appreciate similar ΔPDI/ΔEDI improvement with relief of dyspnea. In the latter position, we found an increase in FRC in normal subjects and in some COPD patients. Thus, probably other factors may have to be taken into account such as a decrease in resistive work and perhaps an increased apposition zone between the diaphragm and the anterior part of the chest wall.

General Muscular Relaxation Techniques?

General relaxation techniques are often proposed to anxious COPD and asthmatic patients. In our experience (unpublished results) even while some objective measurements of relaxation were observed (decreased tonic activity of the chin, decreased Hr) and while most patients experience less dyspnea, no major changes could be found in breathing pattern, the end expiratory level and the use of the respiratory muscles.

External Chest Wall Vibrations?

Finally, we obtained some preliminary results in hyperinflated COPD patients by performing external chest vibrations (unpublished results). We observed a tendency for FRC to decrease while airway resistances increased.

Conclusions

Various techniques may be proposed to patients in order to reduce actively (or passively?) hyperinflation.

Low frequency breathing with active expiration may be one of them though the results indicate some interesting effects but also some deleterious ones, so that it is difficult to be positive about the interest of this technique. Body position, relaxation techniques and external chest vibrations have still to be studied in order to define their relative importance.

References

1. Collett W.P., Engel L.A.: Efficiency of breathing during hyperinflation. In: Grassino A. et. al. (Eds.): *Respiratory muscles in chronic obstructive pulmonary disease*, London, Springer Verlag, 1988, 89-93
2. Tobin M.J.: Respiratory muscles in disease. Clinics in chest medicine, 1988; 9: 263-286
3. Faling L.J.: Pulmonary rehabilitation-Physical modalities. Clinics in chest medicine, 1986; 7: 599-618
4. Sergysels R., Willeput R., Lenders D., Vachaudez J.P., Schandevyl W., Hennebert A.: Low frequency breathing at rest and during exercise in severe chronic obstructive bronchitis. Thorax, 1979; 34: 536-539
5. Sainte-Croix A., Willeput R., Lenders D., Vachaudez J.P., Schandevyl W., Sergysels R.: Implications fonctionnelles du choix du niveau ventilatoire au cours de la ventilation dirigée imposée à des patients atteints de bronchopathie chronique obstructive. Acta Tub. Belg. 1978; 69: 113-126
6. Willeput R., Vachaudez J.P., Lenders D., Nys A., Knoops T., Sergysels R.: Thoracoabdominal motion during physiotherapy in patients affected by chronic obstructive lung disease. Respiration, 1983; 44: 204-214
7. Erpicum B., Willeput R., Sergysels R., De Coster A.:Does abdominal breathing below CRF give a mechanical support for inspiration? Clin. Resp. Physiol. 1984; 20: 117 (abstract)
8. Bellemare F., Grassino A.: Force reserve of the diaphragm in patients with chronic obstructive pulmonary disease. J. Appl. Physiol. 1983; 55: 8-15
9. Dretz S.W., Sharp J.T.: Electrical and mechanical activity of the diaphragm accompanying body position in severe chronic obstructive pulmonary disease. Am. Rev. Resp. Dis. 1982; 125: 275-280

19. Negative Pressure Ventilation as a Treatment Modality in Patients with Hyperinflation

N.AMBROSINO, S.NAVA, C.FRACCHIA, C.RAMPULLA
"S. Maugeri Foundation" Pavia, Care and Research Institute, Medical Center of Rehabilitation, IRCCS, Fondazione Clinica del Lavoro, Montescano, Pavia, Italy

Introduction

Patients suffering from chronic obstructive pulmonary disease (COPD) and hyperinflation often show signs of compromised inspiratory muscle (IM) performance such as dyspnea, tachypnea, paradoxical thoraco-abdominal motion during inspiration, hypoxemia, hypercapnia, reduced vital capacity and exercise performance and decreased of inspiratory muscle strength and endurance.[1,2] All these signs may be due to a condition of chronic IM fatigue which may contribute to the pathogenesis of respiratory failure.[3] Hyperinflation, malnutrition, hypoxemia, hypercapnia and abnormal lung mechanics interact to increase the risk of IM fatigue in COPD patients.[1,2,4-6]

Hyperinflation renders the respiratory muscles prone to fatigue as a result of several pathophysiological factors. Among these there are geometrical changes in the diaphragm,[1,7] the overuse of extradiaphragmatic muscles to meet the ventilatory demands,[8] the waste of the pressure developed by the IM to overcome the "intrinsic PEEP".[9] IM training programs have been proposed in COPD; however, most of them have failed to induce an improvement in the signs of chronic IM fatigue.[10] Recent studies suggest that a period of rest of IM by mechanical ventilation may be useful to help them muscles recover from chronic fatigue.[11,12] Intermittent negative pressure ventilation (INPV) with body ventilators requiring no airway intubation seems to be promising for inducing an improvement in functional reserve of IM, but the experience in this field is still limited. Many problems remain to be elucidated:

1) the real degree of IM rest induced by INPV and its effect on IM function;
2) the effect of INPV on pulmonary circulation;
3) the best schedule of treatment, if any, for short and long term.

Inspiratory Muscle Rest

The real degree of inspiratory muscle rest induced by INPV is still debatable. The classical report of Rochester et al.[13] showed that the electrical activity of the diaphragm was reduced during INPV. On the contrary Rodenstein et al. found that diaphragmatic electromyographic activity increased after 5 minutes of INPV at 15 and at 30 cm H_2O in normal subjects[14] and did not change in COPD patients.[15]

To evaluate some physiological aspects of INPV, we have performed studies in normal subjects.[16] Thoraco-abdominal motion and breathing pattern were obtained from a respiratory inductive plethysmograph (RIP). Transdiaphragmatic pressure (Pdi) was obtained as the difference between gastric (Pga) and esophageal (Pes) pressures, measured with a balloon catheter system connected to pressure transducers. Diaphragmatic electromyographic activity (Edi) was recorded by surface electrodes.

In normal subjects, Pgas, Pes, Pdi, ventilation by Rip [Sum, Abdomen (Abd) and Rib Cage (RC) motion signals] and Edi were obtained in the supine position basally and during INPV at -2, -15, -30 cm H_2O respectively, each period lasting 30 minutes. There was a 15 minute interval with the respirator turned off between periods and at the end of the study. In normal subjects ventilation increased as the negative pressure applied was higher, mainly due to increases in tidal volume (Vt), whereas breathing frequency was maintained fairly constant. Pga changes mainly accounted for the progressive and constant reduction in Pdi observed with the increasing negativity of the applied pressure within the same run. Edi showed a short transient increase after the institution of INPV and a progressive reduction with increasing negativity. Similar Edi observations were obtained in COPD patients.[17]

Haemodynamic Effects

Ventilation with positive pressures was found to affect venous return and cardiac output.[18] Ventilation with an "iron lung" was found to show a linear relationship between the increase in venous pressure and the mean pressure applied to ventilate normal humans.[19] In COPD patients with pulmonary artery hypertension we have performed right heart catheterization during an INPV session by means of a cuirass ventilator.[20] By the percutaneous approach of Seldinger a 7 F Swan-Ganz thermodilution catheter was positioned under fluoroscopy in one of the pulmonary artery branches. Mean pulmonary artery, right atrial and wedge pressure together with the systemic blood pressure obtained from a catheter placed in the radial artery and the ECG were continuously monitored and recorded. Cardiac output was estimated by thermodilution. Hemodynamic measurements were made before every 5 minute interval during 30 minutes of INPV and 5 minutes after the ventilator was stopped.

No changes in mean values of hemodynamics were observed during INPV and in the recovery phase.

Clinical Effects

To verify the clinical effectiveness of INPV in COPD studies were carried out in patients with severe COPD in stable condition[17] who met the following criteria: $PaO_2 < 60$ mmHg; $PaCO_2 > 50$ mmHg; MIP < 60 cmH_2O; one of the clinical signs of IM fatigue: dyspnea, tachypnea, abdominal paradox, chest-abdomen asynchrony.[11] Static lung volumes were measured by means of a body plethysmograph; and dynamic lung volumes by means of a pneumotachograph with volume integrator in the seated posture; blood gas analysis was performed using a blood gas analyzer with arterialized blood obtained from the ear lobe. Respiratory muscle strength was assessed by measuring maximal inspiratory (MIP) and expiratory (MEP) pressures at FRC and TLC respectively according to the method of Black and Hyatt.[21] Patient's subjective evaluation of dyspnea was obtained at rest and after mild effort by a visual analog scale (VAS).[22] Exercise performance was assessed by the 12 min walking distance test (12 mwd).[23] Patients performed INPV using either a cuirass or a pneumowrap ventilator. The ventilators were set to deliver negative pressure between -20 and -35 cmH_2O at a rate slightly greater than the patient's. Patients received INPV 6 hours daily for 5 consecutive days. At the end of the study protocol the basal measurements were repeated, 24 hours after discontinuation of the last INPV session. The results were compared with those of patients in a control group who received conventional physiotherapy. After the period of study there were no significant changes in FRC, FEV_1, FEV_1/FVC ratio and MEP PaO_2. A significant improvement in VC, MIP, $PaCO_2$, VAS and 12 mwd was found in study group. No other significant changes apart from a slight decrease in $PaCO_2$ was observed in control patients. Repeated measurements of the study parameters performed during a 4 month follow-up of patients undergoing periodic INPV sessions according to the basal schedule showed that the benefits observed after the first INPV session disappeared in about 3 weeks. Each of the following monthly INPV sessions induced an improvement in the measured parameters lasting 2 weeks.

Discussion

During INPV in healthy subjects we observed[16] an increase in ventilation related to the degree of negative pressure applied, with an increase in Vt and a constant breathing frequency. This means that the ventilator entrained the subjects breathing frequency.

At a pressure of 30 cm H_2O of INPV, subjects increased their ventilation three-fold. This was accompanied by symptoms of hyperventilation such as finger paresthesia suggesting the presence of hypocapnia. The same observations were made by Rodestein et al., who found that diaphragmatic electromyographic activity increased after 5 minutes of INPV both at -15 and at -30 cm H_2O in normal subjects[14] and did not change in COPD patients.[15] Our observations show that mechanical diaphragmatic activity appears to be reduced under INPV. The classical report of Rochester et al. showed that also the electrical activity of the diaphragm was reduced by INPV.[13]

All the patients studied by us[17] suffered from chronic IM fatigue as indicated by their pattern of breathing, dyspnea, hypercapnia and particularly their reduced MIP. Our own and other studies[12,17] demonstrate that IM strength can be improved by a short-term treatment with INPV by means of cuirass and pneumowrap ventilators, therefore showing that a short-term rest of the IM is effective in temporarily reversing some of the processes leading to impaired IM function in COPD.

Chronic fatigue is one of them and may be induced by conditions such as increased work load, change in geometry, malnutrition, hypoxia, hypercapnia and use atrophy.[1,2-4,9]

Workload hypoxia and malnutrition are not likely to be influenced by INPV; in our study obstruction indices such as FEV_1 and FEV_1 / FVC ratio did not change as for PaO_2. Similarly, the geometrical abnormalities associated with hyperinflation were unchanged by INPV as indicated by no change in FRC, and thus changes in MIP could not be ascribed to changes in muscle length, curvature or mechanical arrangement. It cannot be excluded that the improvement in MIP, VAS and 12 mwd observed in patients submitted to INPV could be ascribed to motivation and learning factors, although in our study no change in these parameters was observed in the control group. MIP is a global index of IM function not specifically applicable to any one IM group.

No relevant side effects were observed during INPV. Furthermore, in the side study to which we refer[20] we observed no change in hemodynamics during cuirass ventilation of 7 patients with pulmonary hypertension, and thus INPV may be a more suitable form of ventilation as compared to intermittent positive ventilation (IPPV) which is known to reduce cardiac output[18] and require airway intubation.

The patients we followed up to 4 months were submitted to periodic treatment with INPV. The physiologic changes observed during the INPV session were highly reproducible, the benefits obtained appearing to last about two weeks. Cropp and Di Marco[12] found that five days after the discontinuation of assisted ventilation the maximal duration time, an index of inspiratory muscle endurance, declined to near baseline values. No information was given on behaviour of MIP and $PaCO_2$.

Few studies are reported on the long-term effect of "chronic" treatment with INPV. The main problem in comparing these studies are the differences in the type of patients and in the protocol followed. Recently, Zibrak et al.[24] studied 20 stable COPD patients submitted to INPV by ponchowrap ventilator for 4-6 months. After the period of daily ventilator use these authors observed no clinically significant improvements in FEV_1, FVC, blood gas determinations, MIP, MEP and exercise duration. Furthemore, 11 out of 20 patients studied dropped out of the study because of inability to tolerate the ventilator, while all but one of the 9 who completed the study were unsatisfied with it, using it for less time than recommended. On the other hand Gutierrez et al.[25] have reported that significant improvements in clinical conditions, arterial blood gases and IM strength were achieved and maintained by ventilating 5 COPD patients for 8 hours once a week. This treatment schedule seems to be promising, not requiring a period of hospitalization, but the number of observations seems too small to reach definitive conclusions.

In conclusion, INPV seems to be effective in reducing IM activity, affording rest and improving functional reserve. INPV does not seem to have any adverse hemodynamic effects. The long-term treament schedule, however, remains to be elucidated.

References

1. Rochester D.F., Braun N.M.T., Arora N.S.: Respiratory muscle strength in chronic obstructive lung disease. Am. Rev. Respir. Dis. 1979; 119 (Part 2: 15: 1-168)
2. Rochester D. F., Braun N.M.T.: Determinants of maximal inspiratory pressure in chronic obstructive pulmonary disease. Am. Rev. Respir. Dis. 1985; 132: 42-47
3. Goldberg P., Roussos C.: Alveolar hypoventilation and respiratory muscle fatigue. In *Current Therapy in Critical Care*. New York, Dekker Inc. 1987, pp 184-188
4. Arora N.S., Rochester D.F.: Respiratory muscle strength and maximal voluntary ventilation in undernourished patients. Am. Rev. Respir. Dis. 1982; 126: 5-8.
5. Arora N.S., Rochester D.F.: Effect of hypoxia and resistive work on diaphragm muscle ATP and glycogen concentration in the dog (Abstract). Clin. Res. 1982; 30: 425
6. Juan G., Calverley P., Talam C., Schnader J., Roussos C.: Effect of CO_2 on diaphragm function in normal human beings. New Engl. J. Med. 1984; 310: 874-879
7. Bellemare F., Grassino A.: Force reserve of the diaphragm in patients with chronic obstructive pulmonary disease. J. Appl. Physiol. 1983; 55: 8-15
8. Farkas G.A., Decramer M., Rochester D.F., De Troyer A.: Contractile properties of intercostal muscles and their functional significance. J. Appl. Physiol. 1985; 59: 528-535
9. Rossi A., Gottfried S.B., Zocchi L., Higgs B.O., Lennox S., Calverley P.M.A., Begin P., Grassino A., Milic-Emili J.: Measurement of static compliance of the total respiratory system in patients with acute respiratory failure during mechanical ventilation: the effect of intrinsic positive end-expiratory pressure. Am. Rev. Respir. Dis. 1985; 131: 72-77
10. Pardy R.L., Leith D.E.: Ventilatory muscle training. In: Roussous C., Macklem P.T.(Eds.): *The Thorax*. New York, M.Dekker Inc., Part B, 1984, 1363-1361
11. Rochester D.F., Martin L.M.: Respiratory muscle rest. In: Roussos C. Macklem P.T. (Eds.): *The Thorax*. New York, M. Dekker Inc. Part B. 1984, pp. 1303-1328

150

12. Cropp A., Di Marco A.F.: Effects of intermittent negative pressure ventilation on respiratory muscle function in patients with severe chronic obstructive pulmonary disease. Am. Rev. Respir. Dis. 1987; 135: 1056-1061

13. Rochester D.F., Braun N.T., Laine S.: Diaphragmatic energy expenditure in chronic respiratory failure. The effect of assisted ventilation with body respirators. Am. J. Med. 1977; 63: 223-231

14. Rodestein D. O., Cuttitta G., Stanescu D.C.: Ventilatory and diaphragmatic EMG, changes during negative-pressure ventilation in healthy subjects. J.Appl. Physiol. 1988; 64: 2272-2278

15. Rodenstein D.O., Stanescu D.C., Cuttitta G., Liistro G., Veriter C.: Ventilatory and diaphragmatic EMG responses to negative pressure ventilation in airflow obstruction. J. Appl. Physiol. 1988; 65: 1621-1626

16. Nava S., Ambrosino N., Zocchi L., Fracchia C., Rampulla C.: Assessment of respiratory muscle rest during negative pressure ventilation. (Abstract). Eur. Respir. J. 1989; (suppl. 5): 3858

17. Ambrosino N., Montagna T., Nava S., Negri A., Zocchi L., Brega S., Fracchia C., Rampulla C.: Short term effect of intermittent negative pressure ventilation in COPD patients with respiratory failure. Eur. Respir. J. 1990; 3 (suppl.10): A.1031

18. Cournand A., Motley H.L., Werko L., Richards D.W.J.: Physiological studies of the effects of intermittent positive pressure breathing on cardiac output in man. Am. J. Physiol. 1948; 23: 944-945

19. Beck G.J., Seanor H.E., Barach A.L., Gates D.: Effects of pressure breathing on venous pressure: a comparative study of positive pressure applied to the upper respiratory passageway and negative pressure to the body of normal individuals. Am. J. Med. Sci. 1952; 224: 169-173

20. Rampulla C., Cobelli F., Ambrosino N., Opasich C., Majani U., Riccardi G., Fracchia C., Zocchi L.: Hemodynamic effect of respiratory muscle rest and inspiratory resistive breathing. Eur. Resp. J. 1988 (Suppl 1): 69S

21. Black L., Hyatt R.: Maximal airway pressures: normal values and relationship to age and sex. Am. Rev. Respir. Dis. 1969; 91: 252

22. Woodcock A., Gross E., Gelleri A., Geddes D.: Effects of dehydrocodeine, alcohol and caffeine on breathlessness and exercise tolerance in COPD patients with normal blood gases. New. Engl. J. Med. 1981; 305: 1611-1616

23. Mc Gavin C.R., Gupta S.P., McHardy G.J.R.: Twelve minute walking test for assessing disability in chronic bronchitis. Br. Med. J. 1976; 1: 822-823

24. Zibrak J.D., Hill N.S., Federman E.C., Kwa S.L., O'Donnell C.: Evaluation of intermittent long-term negative-pressure ventilation in patients with severe chronic obstructive pulmonary disease. Am. Rev. Respir. Dis. 1988; 138: 1515-1518

25. Gutierrez M., Beroiza T., Contreras G., Diaj O., Cruy E., Moreno R.R., Lisboa C.: Weekly cuirass ventilation improves blood gases inspiratory muscle strength in patients with chronic airflow limitation and hypercapnia. Am. Rev. Respir. Dis. 1988; 38: 617-623

20. Exercise Training Strategies for COPD Patients

A. Patessio, F. Ioli, C.F. Donner
"S. Maugeri Foundation" Pavia, Medical Center of Rehabilitation, Fondazione Clinica del Lavoro, Veruno, Novara, Italy

Introduction

One of the most disappointing features of chronic obstructive pulmonary disease (COPD) is the fact that lung function shows a continuous deterioration over many years, leading to overt respiratory failure. Up to now there is no way to arrest the disease in an early stage when the rehabilitation techniques are more likely to induce favourable effects in order to relieve dyspnoea, thus improving the quality of life of the patients.

Nevertheless, a multifaceted approach to this disease can lead to improved well-being and reduced hospitalization.

Since one of the most relevant disabilities of COPD patients is the restriction of physical activity, exercise training has been regarded as a mainstay of rehabilitation programs, even if there is an open debate about the physiologic changes induced and the way to carry it out.

In normal subjects endurance training has been shown to increase in the exercising muscle the number of mitochondria, the capillary density, the mitochondrial enzyme concentration and the glycogen store, decreasing its rate of depletion during acute exercise.[1-3] There is also an increase in maximal exercise ventilation (VEmax) without an increase in maximal voluntary ventilation (MVV).

When the techniques of training are applied to COPD patients, we have to take into account several factors of the disease, other than lung function, that may alter the training response. Duration of illness, cardiac function, bronchoreactivity, nutrition and many other factors lead to the impossibility of having homogeneous populations of patients with comparable and thus predictable responses to training.

In the last 30 years many studies have been carried out to clarify the effects of training in COPD patients. There is a consensus as to the improvement of exercise tolerance and well-being of almost all patients undergoing such physical reconditioning, but many controversies still exist about the causes of this improvement.

One approach could be to consider how exercise training can influence the factors that affect exercise performance in COPD patients: alteration in pulmonary mechanics that leads to a decrease in ventilatory capacity,[4] impairment in pulmonary gas exchange[5-7] with an increase in the ventilatory requirement, respiratory muscle fatigue[8] and an abnormal perception of breathlessness,[9] cor pulmonale.[10]

Ventilatory Limitation

These patients have long been considered to be ventilatory limited and prediction equations have been developed to predict VEmax based on the FEV_1 or other respiratory function variables.[11-13]

The bulk of evidence is that ventilatory limitation is the major factor limiting exercise. However, many of these studies show a wide range of variability between the actual VEmax and the predicted VEmax, even in those patients who exhibit marked airway obstruction. Some patients stop exercising when their predicted VEmax is not reached suggesting that other factors contribute to limit their exercise performance. Moreover, exercise training does not seem to improve ventilatory limitation, because it has no effect on lung mechanics or lung function at rest,[14-15] and so VEmax cannot be increased. As a consequence, relieving ventilatory limitation is a hard task to accomplish once a patient is in a stable condition and under a correct pharmacologic treatment aimed at reversing the bronchoconstriction which may be present.

Ventilatory Requirement

The ventilatory requirement is dictated by the VCO_2 (CO_2 production), the level at which $PaCO_2$ is regulated and the VD/VT ratio (a measure of the inefficiency of gas exchange).

In COPD patients the alterations of the ventilation to perfusion ratio cause an increase in wasted ventilation with a higher VD/VT ratio which leads to more ventilation necessary to keep the $PaCO_2$ at the same level. While it seems unlikely that exercise training has some effect on the VD/VT ratio, it may influence the other two determinants of VE (i.e. $PaCO_2$ and VCO_2). At least in normal subjects[16] it has been demonstrated that the ventilatory requirement can be lowered for work loads engendering metabolic acidosis (i.e. above the anaerobic threshold) as a result of endurance training. The authors pointed out that this effect is obtained through a decrease in the blood level of lactate (a well known effect of endurance training) that leads to lesser additional CO_2 generation via the bicarbonate buffering, which provides more CO_2 to be washed out from the lungs resulting in an increased ventilatory demand. Some preliminary results[17] suggest that also in COPD patients such an effect is likely to be obtained if they are able to develop a substantial metabolic acidosis during exercise[18] and if the training work load is high enough. The ability to have an early elevation[19] in blood lactate, greater than in normal subjects for the same work loads, but as in patients with cardiovascular disease, is not linked to the baseline pulmonary function and can be an indirect index of pulmonary vascular disease. Other possible determinants of ventilation during exercise may be epinephrine and norepinephrine and while it cannot be ruled out that they play a minor role in normal subjects,[16] it is still unclear whether they may be important in COPD patients.

Respiratory Muscle Fatigue

The respiratory muscles in COPD patients work under adverse conditions: due to hyperinflation the diaphragm[20] is flat and has to act in the less favourable portion of the length-tension relationship. Moreover, factors such as airway obstruction and hypoxemia further impair their efficiency increasing the work of breathing and decreasing the energy supply.[21] It is probable then that respiratory muscle fatigue or weakness can play a role in limiting exercise.[22]

A specific training of the respiratory muscles in order to improve their strength and endurance[23-25] seems a reasonable way to increase the ventilatory capacity and thus exercise performance, but when considering both the methods used to train the respiratory muscle (resistive breathing and hyperpnea training) up to now controversial results have been obtained.[26,27]

In fact some authors[26] have found an increase in exercise endurance, whereas others[28] have found only an increase in ventilatory endurance.

Dyspnoea

Many factors contribute to an increase in perception of breathlessness during exercise. The quantitative disturbance of the sense of inspiratory loads and the disproportionate ventilation relative to work load are more likely to be the major mechanisms responsible for increased dyspnea. There is a contrast between how much air needs to be ventilated to sustain the metabolic demands and how much air can be breathed: in fact, many COPD patients do not hyperventilate at the maximal work loads and, unlike normal subjects, the respiratory alkalosis in response to the metabolic acidosis is not seen.

The effects of exercise training range from a reduction in breathlessness to a better sense of well-being.[29,30]

Agle et al.[31] suggested that these improvements were due to the reassuring role of medical personnel acting as a desensitizing form of behaviour therapy. Surely it is important to achieve such an effect, because it can lead to increased exercise endurance, even if it is not measurable as a physiologic change.

Desaturation during exercise also plays a role in increasing dyspnea and supplemental oxygen in these cases is helpful, allowing these patients to increase their exercise tolerance and decreasing breathlessness.

Pulmonary Vascular Disease

An early increase in pulmonary artery pressure during exercise is often found in COPD patients, even in those with normal resting values. The response of PAP to exercise is different from that of normal subjects, where it does not increase until cardiac output increases by 3 times its resting value[32] and right ventricular dysfunction has been demonstrated to occur[33] by radionuclide angiography. Whether pulmonary vascular disease can be considered a factor directly limiting exercise performance is still under debate.

The effects of training on pulmonary hemodynamics have yet to be completely elucidated. Small numbers of patients and different training protocols involving too low work loads are the major problems of the published studies. Alpert et al.[34] and Degré et al.[35] demonstrated that there is no change in pulmonary artery pressure after training, whereas the arterio-venous difference increased in one study[35] (as it does in normal subjects) and did just the opposite in the other.[34]

Methods of Training and Evaluation of the Results

The three main types of exercise used for training are walking, walking on a treadmill and cycling. Less often used is arm exercise, which was found to lead to dyssynchronous movements of the abdomen and chest wall,[36] thus interfering with the appropriate action of the respiratory muscle. Walking is the most natural type of exercise, but many people are also familiar with cycling which is, at least in some countries, a very popular hobby.

The way of assessing exercise performance and of carrying out training programs should be taken into account when evaluating the results obtained. In fact, the cycle ergometer is more useful in evaluating physiologic changes after training because the mechanical efficiency of exercise varies little and so the work load performed is easily reproducible and measurable.[37] On the other hand, familiarization with treadmill exercise can lead to improvement of skill in performing the exercise.[38]

The use of these devices in assessing exercise tolerance usually implies a well equipped laboratory, where it is possible to measure the physiologic variables of exercise. A simpler tool to quantify easier exercise tolerance and to evaluate programmes based on walking is the 12 minute walking test, which was originally described by Cooper[39] who evaluated 115 military personnel, showing that the distance covered in the 12 minute field performance test correlated well (r=0.897) with maximum VO_2 assessed during an incremental treadmill test. The more standardized 12 minute test in the evaluation of patients suffering from chronic bronchitis was introduced by McGavin et al.,[40] who found a weaker relationship between the distance covered and VO_2max on the cycle ergometer (r=0.53). In fact, many factors other than physiologic respiratory and cardiovascular functions play a role: motivation, attitude, familiarity with the procedure, self paced nature of the test and neuromuscular function. So the results of training programmes expressed as improvement of the distance walked are useful in reflecting a better ability to perform everyday activities, but not in clarifying physiologic mechanisms of training activity.[41]

Another important problem which has yet to be clarified is the use of supplemental oxygen: the validity of long-term oxygen therapy in reversing the progression of pulmonary hypertension has been recently proven[42-43] and reviewed.[44] In addition administering oxygen during exercise in COPD patients improves their exercise tolerance, but it is still unclear which patients would benefit from oxygen during exercise (decrease in breathlessness while exercising has also been demonstrated in patients with normoxemia)[45] and whether supplemental oxygen is useful in ameliorating the effects of exercise training.

Conclusion

Exercise training can be considered an important and simple way to improve the quality of life of patients suffering from chronic airway obstruction. When correctly prescribed it is safe and commonly well accepted by patients. However, despite the large number of studies, more investigations are needed to clarify the physiologic mechanisms involved in obtaining such results and to better define which patients are more likely to benefit from training and the way to carry it out.

References

1. Baldwin F.M., Fitts R.H., Booth F.W., Winder W.W., Holloszy J.O. : Depletion of muscle and liver glycogen during exercise: protective effect of training. Pfluegers Arch. 1975; 354: 203-212
2. Saltin B., Gollnick P.D. : Skeletal muscle adaptability: significance for metabolism and performance. In: *Handbook of Physiology. Skeletal muscle*. Washington. Am. Physiol Soc, 1983; sect 10, chapt 19, pp.555-631

3. Baldwin K.M., Winder W.W., Holloszy J.O.: Respiratory capacity of white, red and intermediate muscle: adaptive response to exercise. Am. J. Physiol. 1973; 222: 373-378

4. Clark T.J.H., Freedman S., Campbell E.J.M.: The ventilatory capacity of patients with chronic airways obstruction. Clin. Sci. 1969; 36: 307-316

5. Jones N.L. Pulmonary gas exchange during exercise in patients with chronic airway obstruction. Clin. Sci. 1966; 31: 39-50

6. Jones N.L., McHardy G.J.R., Naimark A.: Physiological dead space and alveolar-arterial gas pressure differences during exercise. Clin. Sci. 1966; 31: 19-29

7. Minh V.D., Lee H.M., Dolan G.: Hypoxemia during exercise in patients with chronic obstructive pulmonary disease. Am. Rev. Respir. Dis. 1979; 120: 787-794

8. Bye P.TP., Farkas G.A., Roussos C.: Respiratory factors limiting exercise. Ann. Rev. Physiol. 1983; 45: 439-451

9. Gandevia S.C., Killian K.J., Campbell E.J.M.: The effect of respiratory muscle fatigue on respiratory sensations. Clin. Sci. 1981; 60: 463-466

10. Matthay R.A., Berger H.J.: Cardiovascular function in cor pulmonale. Clin. Chest. Med. 1983; 4: 269-295

11. Dillard T.A., Piantadosi S., Rajagopal K.R.: Prediction of ventilation at maximal exercise in chronic airflow obstruction. Am. Rev. Respir. Dis. 1985; 132: 230-235

12 Spiro S.G. : Exercise testing in clinical medicine. Br. J. Dis. Chest. 1977; 71: 145-172

13. Carter R., Pevler M., Zinkkgraf S., Williams J., Fields S.: Predicting maximal exercise ventilation in patients with chronic obstructive pulmonary disease. Chest 1987; 92: 253-259

14. Lertzman M.M., Cherniack R.M.: Rehabilitation of patients with chronic obstructive pulmonary disease. Am. Rev. Respir. Dis. 1976; 114: 1145-65

15. Belman M.J., Wasserman K.: Exercise training and testing in patients with chronic obstructive pulmonary disease. Basics of RD, 1981, 10: 1-6

16. Casaburi R., Storer T.W., Wassermann K.: Mediation of reduced ventilatory response to exercise. J. Appl. Physiol. 1987; 63: 1533-1538

17. Casaburi R., Wasserman K., Patessio A., Ioli F., Zanaboni S., Donner C.F.: A new perspective in pulmonary rehabilitation: anaerobic threshold as a discriminant in training. Eur. Respir. J. 1989 (In press)

18. Sue D.Y., Wasserman K., Moricca R.B., Casaburi R.: Metabolic acidosis during exercise in patients with chronic obstructive pulmonary disease. Chest. 1988; 94: 931-938

19. Shuey, C. B. Jr., Pierce A. K, Johnson R. L. Jr.: An evaluation of exercise tests in chronic obstructive lung disease. J. Appl. Physiol. 1969; 27 (2): 256-261

20. Roussos C., Macklem P.T.: Diaphragmatic fatigue in man. J. Appl. Physiol. 1977; 43: 189-197

21. Field S., Kelly S.M., Macklem P.T.: The oxygen cost of breathing in patients with cardiorespiratory disease. Am. Rev. Respir. Dis. 1982; 126: 9-13

22. Grassino A., Gross D., Macklem P.T. et al.: Inspiratory muscle fatigue as a factor limiting exercise. Bull. Eur. Physiopathol. Respir. 1979; 15: 105-111

23. Pardy R.L., Rivington R.N., Despas P.S., Macklem P.T.: The effects of respiratory muscle training on exercise performance in chronic airflow limitation. Am. Rev. Respir. Dis. 1981; 123: 426-433

24. Peress L., McClean P., Woolf C., Zamel N.: Respiratory muscle training in severe chronic obstructive pulmonary disease. Am. Rev. Respir. Dis. 1979; 119 (part 2): 157

25. Andersen J.B., Dragsted L., Kann T. :Resistive breathing training in severe chronic obstructive pulmonary disease. A pilot study. Scand. J. Respir. Dis. 1979; 60 (3): 151-155

26. Belman M.J., Mittman C.: Ventilatory muscle training improves exercise capacity in chronic obstructive pulmonary disease patients. Am. Rev. Respir. Dis. 1980; 121, 273-280

27. Bjerre-Jepsen K., Scher N.H., Koh-Jensen A.: Inspiratory resistance training in severe chronic obstructive pulmonary disease. Eur. J. Respir. Dis. 1981; 62: 404-411

28. Levine S., Weiser P., Gillen J.: Evaluation of a ventilatory muscle endurance training program in the rehabilitation of patients with obstructive pulmonary disease. Am. Rev. Respir. Dis. 1986; 133: 400-406

29. Cockcroft A.E., Sanders M.J., Berry G.: Randomized controlled trial of rehabilitation in chronic respiratory disability. Thorax 1981; 36: 200-203

30. Sinclair D.J.M., Ingram C.G.: Controlled trial of supervised exercise training in chronic bronchitis. Br. Med. J. 1980; 1: 519-521

31. Agle D.P., Baum G.L., Chester E.H. et al.: Multidiscipline treatment of chronic pulmonary insufficiency: functional status at one year follow-up. In: Johnson R.F. (Ed.). *Pulmonary Medicine*: A Hahnemann Symposium. New York, Grune and Stratton, 1973; p. 355

32. Robin E.D., Gaudio R.: Cor pulmonale; D.M. May 1970, pp. 1-38

33. Matthay R.A., Berger H.J., Davies R.A. et al.: Right and left ventricular exercise performance in chronic obstructive pulmonary disease: radionuclide assessment. Ann. Intern. Med. 1979; 93: 234-239

34. Alpert J.B., Bass H., Szucs M.M., Banas J.S., Dalen J.A., Dexter L.: Effects of physical training on hemodynamics and pulmonary function at rest and during exercise in patients with chronic obstructive pulmonary disease. Chest 1974; 66: 647-651

35. Degré S., Sergysels R., Mesin R., Vandermoten P., Salhadin P., Denolin S.H., De Coster A. Hemodynamic responses to physical training in patients with chronic lung disease. Am. Rev. Respir. Dis. 1974; 110: 395-402

36. Celli B.R., Rassulo J., Make B.J. :Dyssynchronous breathing during arm but not leg exercise in patients with chronic airflow obstruction. New Engl. J. Med. 1986; 314: 1485-1490

37. Hansen J.E., Casaburi R., Cooper D.M., Wasserman K. Oxygen uptake as related to work rate increment during cycle ergometer exercise. Europ. J. Appl. Physiol. 1988; 57: 140-145

38. Nicholas J.J., Gilbert R., Gabe R., Auchincloss J.H.: Evaluation of an exercise therapy program for chronic obstructive pulmonary disease. Am. Rev. Resp. Dis. 1970; 102: 1-9

39. Cooper K.H.: A means of assessing maximal oxygen intake. JAMA 1968; 203: 135-138

40. McGavin C.R., Gupta S.P., McHardy G.J.R.: Twelve-minute walking test for assessing disability in chronic bronchitis. Br. Med. J. 1976; 1: 822-823

41. Donner C.F., Patessio A.: Performance indicators in chronic obstructive lung disease: walking test. Eur. Respir. J. 1989 (In press)

42. Timms R.M., Khaja F.U., Williams G.W.: Hemodynamic response to oxygen therapy in chronic pulmonary disease. Ann. Intern. Med. 1985; 102: 29-36

43. Weitzenblum E., Sautejeau A., Ehrahrt M., Mammosser M., Pelletier A.: Long term oxygen therapy can reverse the progression of pulmonary hypertension in patients with chronic obstructive pulmonary disease. Am. Rev. Respir. Dis. 1985; 131: 493-498

44. Levil-Valensi P., Aubry P., Donner C.F., Robert D., Ruhle K.H., Weitzenblum E., Wurtemberger R.: Recommendations for long term oxygen therapy. Eur. Respir. J. 1989; 2: 160-164

45. Woodcock A.A., Gross E.R., Geddes D.M.: Oxygen relieves breathlessness in pink puffers. Lancet 1981; 1: 907-909

21. Respiratory Muscle Training in Health and Disease

D. Gross[1], A. Appelbaum[2]
1. Department of Anesthesiology, Hadassah University Hospital, Jerusalem, Israel
2. Department of Cardiothoracic Surgery, Hadassah University Hospital, Jerusalem, Israel

Respiratory muscles, like other skeletal muscles, are subject to fatigue.[1-2] This fatigue can be diagnosed objectively by observing changes in the frequency spectrum of the EMG.[3] When fatigue of the respiratory muscles occurs, they can fail in their role as a pressure generator and the resulting effects upon gas exchange are potentially grave.[4]

In respiratory muscles as in other skeletal muscles, if trained for strength and endurance, their fatigue can be prevented or its onset can be delayed.

The idea of ventilatory exercises had already been put forward by the Chinese three thousand years ago.[5] In the last century several devices were designed without a good understanding of respiratory physiology. For example, the device developed by Ruebsam in 1902 trained the diaphragm as an expiratory muscle and not as an inspiratory one.[6] Recently, it has been demonstrated that appropriate training of the ventilatory muscles in normal subjects increases strength and endurance.[7] Leith and Bradley measured their normal subjects hyperventilating at the highest level of ventilation as a percentage of their maximum voluntary ventilation (VE/MVV).[7] An improvement in VE/MVV from 85% to 95% was observed.[7] Their training program, however, is cumbersome and can only be done in a hospital; hence patients are less compliant. Most of the work on the effect of respiratory muscle training was done after this study by Leith and Bradley had been published in 1976. Endurance training by hyperventilation, the technique used by Leith and Bradley, and upper body exercise were demonstrated to increase respiratory muscle endurance in patients with cystic fibrosis.[8]

In 1980 we studied tetraplegic patients, who lose about 60% of their inspiratory muscle force with a greater loss of their expiratory force. These patients were trained with an inspiratory resistance, the smallest that produced electromyographic (EMG) changes indicative of fatigue,[3] 15 minutes twice daily six days a week for 16 weeks. A significant increase in maximal mouth pressure (Pmmax) from 64 to 85 mmHg. Inspiratory endurance was depicted by the critical pressure (Pmcrit), namely, the minimum level of mouth pressure developed during inspiration, which produced the EMG changes indicative of fatigue.[9] The Pmcrit of quadriplegic patients as percentage of Pmmax was 11-15.5% compared to normal subjects whose Pmcrit is 50-70% Pmmax. It was noticed that after a few weeks of training inspiratory resistance no longer elicited an EMG patern of fatigue. Significant increases in maximal mouth pressure and in tolerance to higher inspiratory resistance were found in all patients; hence Pmcrit increase indicated an elevation in respiratory muscle endurance (Fig. 1).

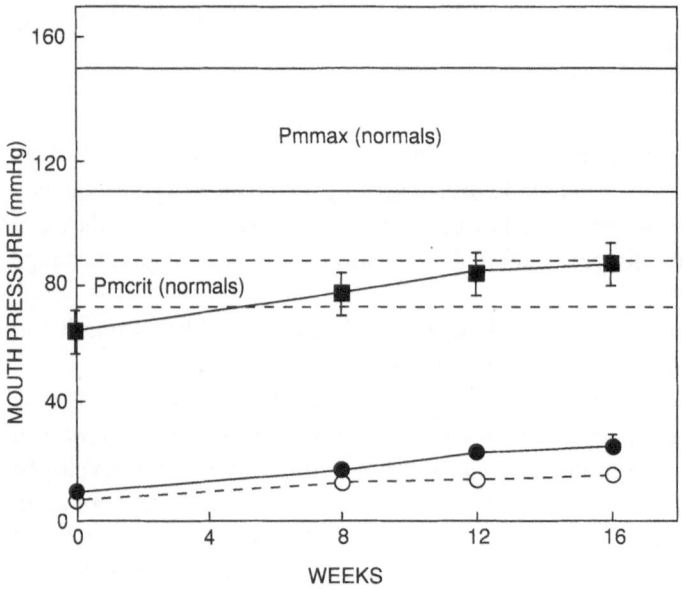

Fig. 1. Changes in mouth pressure (Pm) during maximum effort and critical pressure (Pmcrit) for normal subjects (shaded area) and quadriplegic patients (graphs). Pmcrit is the value between Pm inducing fatigue (−) and that which does not induce fatigue (o). Pmmax for quadriplegic patients (•)).

Improvement in strength and ventilatory capacity of acute quadriplegic patients, two weeks post injury has been demonstrated.[10] In our personal experience, ventilator dependent patients have also been shown to benefit from this treatment. It has been shown that the inspiratory muscles of patients with chronic airflow limitation are weaker than those of normal persons.[11] Pardy et al.[12] using the method of Gross et al.[9] trained COPD patients for 8 weeks and found no changes in maximal inspiratory force and functional residual capacity (FRC). However, patients increased their endurance time to exercise and increased the distance that they could walk in 12 min. As in the study by Gross et al.[9] they found the reversal of EMG pattern of fatigue after 8 weeks of resistance training, indicating an improvement in respiratory muscle endurance. The same authors showed in another study[13] that a simple home training program of the inspiratory muscles was more effective than physiotherapy in improving exercise performance of some patients with severe chronic airflow limitation. Resistive breathing in severe chronic obstructive pulmonary disease has also been shown to induce ventilatory improvement.[14] That is, higher resistance could be tolerated without signs of fatigue and it helped them in their daily living in that they were able to do more without getting short of breath.

Some COPD patients show no response to training, or the degree of response is variable and possibly reflects the intensity of the training program used.[15] The major benefit that should be expected from a respiratory muscle training program is an improvement in respiratory muscle performance. Some studies done on COPD patients have demonstrated improvement in exercise performance while others have not. Conflicting results are found. For example, the first study by Belman and Mittman[16] demonstrated a significant enhancement of ventilatory and exercise capacity while another study by Belman et al. found no improvement.[17] These conflicting results could be due to two major reasons:

1. Reasons related to the lack of a true training effect on the respiratory muscles.

2. Reasons related to the fact that exercise limitation in COPD is multifactorial and the outcome depends on the type of patients included in the study.

Estrup et al.[18] have demonstrated that respiratory muscle training by resistance breathing, using the method of Gross et al.[9] increased vital capacity significantly and respiratory muscle endurance in patients with progressive muscular dystrophy and atrophia spinales types II and III.

An advance in the field of resistive breathing devices was the flow directed or incentive spirometer resistive breathing device which has also been demonstrated to improve respiratory muscle strength and endurance as well as exercise capacity.[19] It can be concluded from these studies that resistive breathing is an effective method of improving the strength and endurance of the respiratory muscles in most patients.

Prophylactic Ventilatory Treatment for Open Heart Surgery

Pulmonary complications after surgery are a leading cause of postoperative morbidity and mortality.[20,21] There is general agreement that cardiac, thoracic and upper abdominal surgery carries the highest risk of postoperative respiratory complications (greater than 20%).[21] The incidence of pulmonary complications is greater in patients with COPD and it is greater in patients undergoing thoracic or upper abdominal surgery than lower abdominal or peripheral surgery.[22,23] This might be explained by marked alterations in diaphragmatic function after thoracic and/or upper abdominal surgery.[24] Postoperative pulmonary complications could be due to either pump related or lung related problems inducing pump related complications as a secondary cause. Pump related complications consist in inefficient ventilatory mechanics, such as low compliance, incoordinated breathing, ventilatory muscle fatigue, muscle weakness, phrenic paralysis and others, normally manifested by elevated arterial PCO_2. Lung related problems are pulmonary emboli, ARDS and those of gas exchange, normally manifested by hypoxia. These would induce an elevated respiratory load on the one hand and inefficient ventilatory muscle work on the other, which could develop into respiratory muscle fatigue and failure.[25-26]

Both ARDS and gas exchange problems reduce the amount of energy supplied to the ventilatory muscles. The reduced energy supply to the muscles combined with the elevated load and work of breathing will result in reduced efficiency of the respiratory muscles, and consequently, their degree of fatigue will be increased.[26,27]

It was reasonable to assume that strengthening and increasing the endurance of the ventilatory muscles before surgery could prevent post-surgical complications because the system will be prepared to withstand the elevated post-surgical load and decreased efficiency due to the effect of anesthesia and surgical trauma.

It was proposed to investigate a prophylactic respiratory treatment for open heart surgery by training the ventilatory muscles for one month prior to surgery, utilizing the resistive breathing training technique.[9]

Subjects

For this study 100 patients were divided into two groups; group I (n=50) (untreated group) patients were evaluted but not treated; group II (n=50) (treated group) patients were evaluated and treated using resistive breathing training for one month at home. 80% of the patients, in each group, underwent coronary artery bypass graft (CABG) and 20% had valve(s) fixed or replaced.

Methods

Pulmonary function tests (PFT) using the vitalograph compact spirometer (Vitalograph).

Ventilatory muscle endurance test: period over which patients could maintain resistive breathing with a constant flow rate.

Post-surgical Monitoring

The following parameters were monitored: chest X-ray, PaO_2, $PaCO_2$, duration of mechanical ventilation, postoperative complications.

Procedures

Patients underwent a complete evaluation one month prior to surgery, at which time the resistive breathing device was given to them (inspiratory muscle trainer, DHD Medical product, U.S.A.). They were instructed to breathe with it for 10 minutes three times every day throughout the whole month. The tests were repeated when patients arrived in the hospital prior to surgery (one month after the last evaluation). Monitoring of these patients was performed immediately after surgery and continued daily for 7 days. Complications were indicated and in a follow-up visit in the clinic one month post surgery a complete evaluation was performed.

Results

Forced vital capacity (FVC), one month before surgery was $83.8 \pm 2.9\%$ and $86.4 \pm 2.7\%$ predicted for the untreated and the treated groups respectively (Fig. 2). Two days before surgery, in the untreated group FVC did not change while it increased to $94.0 \pm 2.3\%$ (p<.04) in the treated group. The higher value in the treated group remained for one month post surgery (Fig. 2).

Maximum voluntary ventilation (MVV) was the same for the two groups one month prior to surgery (MVV=81.8 ± 4.1 and $74.4 \pm 3.3\%$ of predicted value for the treated and untreated groups).

MVV increased from 74.4 ± 3.3 to $92.8 \pm 6.1\%$ (p< .001) in the treated group but remained unchanged in the untreated group (Fig. 3).

Resistive breathing duration increased from 4.8 ± 0.3 minutes to 10.9 ± 0.3 minutes (p< .001), while in group I (untreated) it was 5.7 ± 0.4 minutes and did not change (Fig. 4).

In the post-surgical recovery period, the duration of postoperative mechanical ventilation decreased from 23.8 ± 1.7 hours in the untreated group to 16.1 ± 1.0 hours (p< .0003) in the treated group (Table I).

Table I. Duration of postoperative mechanical ventilation in the untreated and treated groups.

	Untreated	Treated	Significance
$PaCO_2$	47.7 ± 2	$40.7 \pm .5$	p<.0002
PaO_2	93.5 ± 9.6	109 ± 14.5	ns
Duration of intubation	23.9 ± 1.7	16.1 ± 1	p<.0003
Complications	27%	4%	p<.0002

$PaCO_2$, two hours post successful extubation, was found to be 47.7 ± 2.0 torr in the untreated group and 40.7 ± 0.5 torr in the treated group (p< .0002), PaO_2 was slightly higher but not significantly higher in the group using resistive breathing training compared with the untrained group. Severe post-surgical complications leading to prolonged mechanical ventilation, CPAP treatment or prolonged oxygen

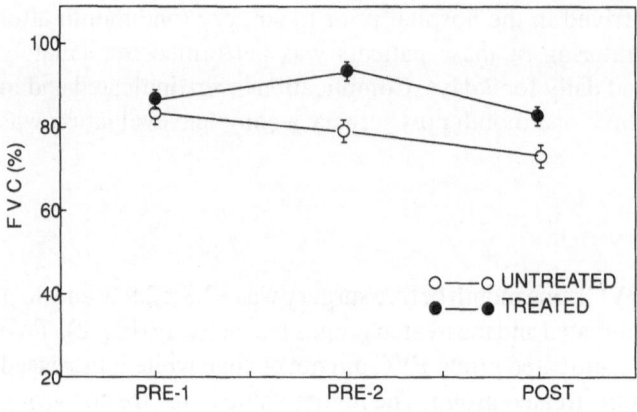

Fig. 2. Forced Vital Capacity (FVC) measured one month before, two days before and one month after open heart surgery in the untreated (o) and treated (•) groups.

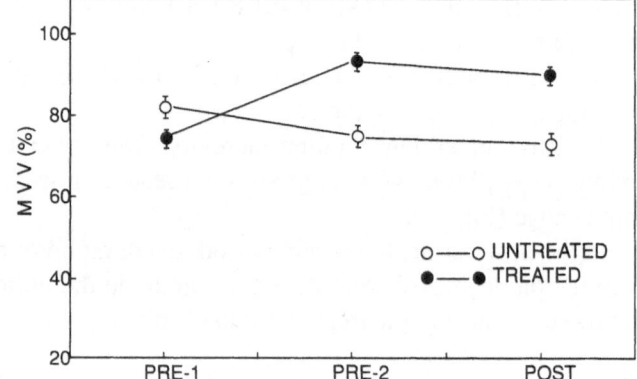

Fig. 3. Maximum Voluntary Ventilation (MVV) measured one month before, two days before and one month after open heart surgery in the untreated (o) and treated (•) groups.

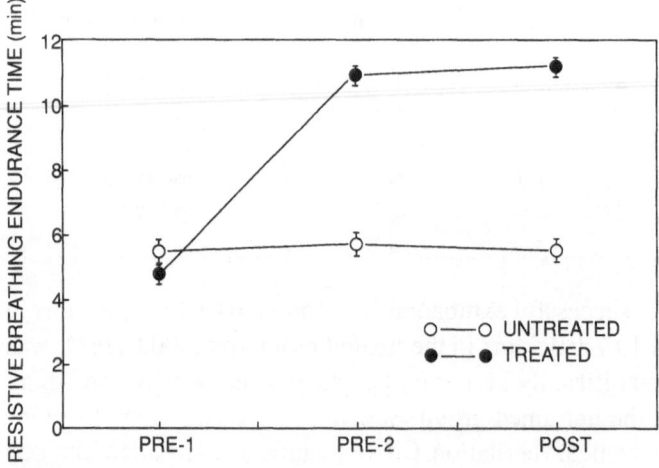

Fig. 4. Endurance time of resistive breathing one month before, two days before and one month after open heart surgery in the untreated (o) and treated (•) groups.

maintenance decreased from 27% to 4% ($p < .0002$) as a result of the resistive breathing training for one month before surgery.

Discussion

The main results of this study were that resistive breathing training in cardiac patients improved the pulmonary functions and ventilatory endurance capacity before surgery, resulted in a better recovery, better blood gases, reduced duration of postoperative mechanical ventilation and reduced post-surgical complications.

Cardiac patients have an inefficient perfusion due to their inefficient heart and narrowed vessels. Therefore, less energy is supplied to the ventilatory muscles.[28] Blood flow to the ventilatory muscles is normally elevated with increased ventilation.[29] However, it can be limited by lower perfusion as occurs in cardiac patients, in the case of tamponade, hemorrhage[30] or by very strong muscle contraction. A very strong contraction will close the muscle capillary bed. This might occur during very high pulmonary load. Bellemare et al.[29] demonstrated that in normal dogs low force contraction increased blood flow, while strong contractions will result in limitation of blood flow during contraction (inspiration) and hyperemia during relaxation (expiration). This may not be the case in cardiac patients whose peripheral perfusion is not as efficient. However, training improves the efficiency of the respiratory muscles to work and thus the decreased blood flow may not have such a severe effect on these patients.

The anesthesia during surgery may result in increased postoperative secretions and the large amounts of fluids given during surgery and the surgical trauma with the effect of the bypass combined with an oxygenator may induce permeability defects.

Together they may cause increased secretions during the initial postoperative period as well as perhaps influx of fluid into the lungs. The increased secretions and accumulation of fluid in the lung will induce elevation of the respiratory load. Patients normally respond by rapid shallow breathing which is inefficient. This inefficient work of breathing is further amplified by the post-surgical pain. Strengthening the inspiratory muscles therefore can lead to deeper and more relaxed breathing, preventing some further complications, such as atelectasis. The ability to breathe more deeply also enables the patients to cough more effectively and thus get rid of the secretions. This study demonstrates once more that improved respiratory muscle strength and endurance by ventilatory muscle training can modify the clinical course of the traumatic post-surgical respiratory system.

The role of the respiratory muscles has been compared to that of the heart muscle in the cardiovascular system.[31,32] Both have a pumping action and must contract continuously throughout life without protracted rest. This is particularly true with respect to the principal inspiratory muscle, the diaphragm.

Endurance training requires exercise of sufficient load, speed and duration in order to induce the enzymatic system of the citric acid cycle and the electron transport system to be overloaded. This is acheived by a very low resistive load.[33] Thus, patients do not have to feel difficulty performing this treatment.

The device is small, and can thus be carried in the pocket or in a small bag, and the training can be done at home. A conceptual scheme for respiratory muscle training is presented in Fig. 5.

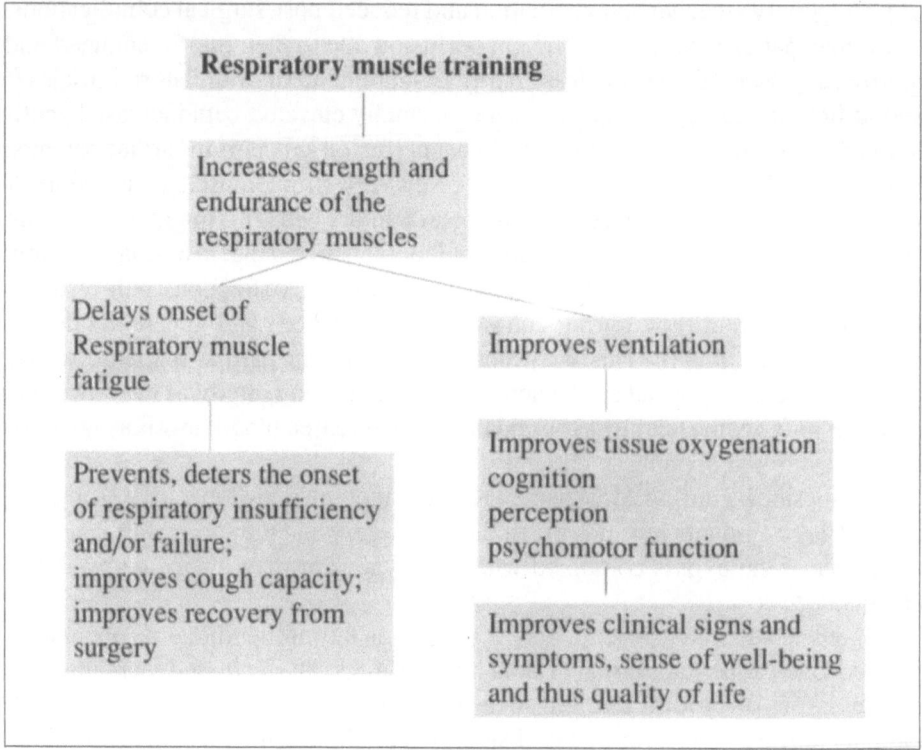

Fig. 5. Conceptual scheme of the effect of respiratory muscle training.

Respiratory muscle training increases the strength and/or endurance of the ventilatory muscles. It can affect the patient in two ways: first in the acute situation, for example post open heart surgery when the onset of ventilatory muscle fatigue needs to be deterred so that respiratory insufficiency or failure can be prevented. The second effect is of a long term nature; it improves ventilation as a result of increased strength and endurance, decreases the work of breathing and/or improves tissue oxygenation. These factors play an important role in the clinical course of the patient's activity and the quality of daily living.

The elevated respiratory muscle efficiency can also be manifested in the patient's cognition, perception and psychomotor function. Therefore, it affects the patient's sense of well-being and quality of life.

References

1. Roussos C., Macklem P.T.: Diaphragmatic fatigue in man. J. Appl. Physiol. 1977; 43: 189-197.
2. Roussos C., Fixley M., Gross D., Macklem P.T.: Fatigue of the inspiratory muscles and their synergic behavior. J. Appl. Physiol. 1979; 46(5): 897-904.
3. Gross D., Grassino A., Ross W.R.D., Macklem P.T.: Electromyogram pattern of diaphragmatic fatigue. J. Appl. Physiol. 1979; 46: 1-7.
4. Cohen C., Zagelbaum G., Gross D., Roussos C., Macklem P.T.: Clinical manifestations of inspiratory muscle fatigue. Am. J. Med. 1982; 73: 308-316.
5. Lyons A.S., Petrucelli R.J.: Medicine, An illustrated history. New York, H.N. Adams Inc. 1978, pp. 121-124.
6. Belman M.J.: Clinics in chest medicine, New York, W.B. Saunders Comp. 1988, p. 288.
7. Leith D., Bradley M.: Ventilatory muscle strength and endurance training. J. Appl. Physiol. 1976; 41: 508-516.
8. Keens T.G., Krastins I.R.B., Wannamaker E.M.: Ventilatory muscle endurance training in normal subjects and patients with cystic fibrosis. Am. Rev. Resp. Dis. 1977; 116: 843-860.
9. Gross D., Ladd H.W., Riley E.J., Macklem P.T., Grassino A.: The effect of training on strength and endurance of the diaphragm in quadriplegia. Am. J. Med. 1980; 68: 27-35.
10. Hornstein S., Ledsome J.R.: Ventilatory muscle training in acute quadriplegia. Physiotherapy Canada.
11. Rocherster D.F., Braun N.M.T., Arora N.S.: Respiratory muscle strength in chronic obstructive pulmonary disease. Am. Rev. Resp. Dis. 1979; 119, 2: 151-154.
12. Pardy R.L., Rivington R.N., Despas P.J., Macklem P.T.: The effect of inspiratory muscle training on exercise performance in chronic airflow limitation. Am. Rev. Resp. Dis. 1981; 123: 426-434.
13. Pardy R.L., Rivington R.N., Despas P.J., Macklem P.T.: Inspiratory muscle training compared with physiotherapy in patients with chronic airflow limitation. Am. Rev. Resp. Dis. 1981; 123: 421-426.
14. Andersen J.B., Dragstem L., Kann T., Johansen S.H., Nielsen K.B., Karbo E., Bentzen L.: Respiration. J. Respir. Dis. 1979. 60: 151-156.
15. Pardy R.L., Reid D.W., Belman M.J.: Respiratory muscle training. In: M.J. Belman (Ed.) *Clinics in chest medicine*, New York, W.B. Saunders Comp. 1988; pp. 287-296.
16. Belman M.J., Mittman C.: Ventilatory muscle training improves exercise capacity in chronic obstructive pulmonary disease patients. Am. Rev. Resp. Dis. 1980; 121: 273-280.
17. Belman M.J., Thomas S.G., Lewis M.I.: Resistive breathing training in patients with chronic obstructive pulmonary disease. Chest 1986; 90 (5): 662-669.
18. Estrup C., Lyager S., Noeraa N., Olsen C.: Effect of respiratory muscle training in patients with neuromuscular diseases and in normals. Respiration 1986; 50: 36-43.
19. Larson M., Kim M.J., Fann D.: Respiratory muscle training with the incentive spirometer resistive breathing device. Heart and Lung 1984; 13(4): 341-345.

20. Gass D.G., Olsen G.N.: Preoperative pulmonary function to the predict post operative morbidity and mortality. Chest 1986; 89(1): 127-135.

21. Laszlo G., Archer G.G., Darrell J.H., Dawson J.M., Fletcher C.M.: The diagnosis and prophylaxis of pulmonary complications of surgical operation. Brit. J. Surg. 1973; 602: 129-134.

22. Stein M., Cassara G.L.: Preoperative pulmonary evaluation and therapy for surgery patients. J. Am. Med. Assoc. 1970; 211: 787-790.

23. Ford G.T., Whitelaw W.A., Rosenal T.W., Cruse P.J., Gunther C.A.: Diaphragm function after abdominal surgery in humans. Am. Rev. Resp. Dis. 1983; 127: 431-436.

24. Latimer R.G., Dickman M., Day W.C., Cunn M.L., Schmidt D.C.: Ventilatory pattern and pulmonary complications after upper abdominal surgery determined by preoperative and postoperative computerized spirometry and blood gas analysis. Am. J. Surg. 1972; 122: 622-623.

25. Derenne J.P.H., Macklem P.T., Roussos C.: The respiratory muscles: mechanics, control and pathophysiology. Am. Rev. Resp. Dis. 1978; 118: 581-601.

26. Field S., Kelly S.M., Macklem P.T.: The oxygen cost of breathing in patients with cardiorespiratory disease. Am. Rev. Resp. Dis. 1982; 126: 9-13.

27. Bryant S., Edwards R.H.T., Faulkner J.A., Hughs R.L., Roussos C.: Respiratory muscle failure: fatigue or weakness. Chest 1986; 89: 116-124.

28. Monod H., Scherrer J.: The work capacity of a synergic muscular groups. Ergonomics 1965; 8: 329-337.

29. Bellemare F., Wight D., Lavigne C., Grassino A.: Effect of tension and timing of contraction on blood flow of the diaphragm. J. Appl. Physiol. 1983; 54: 1597-1606.

30. Viires N., Silly M., Aubier M., Rassidakis A., Roussos C.: Effect of mechanical ventilation on respiratory muscles and regional blood flow distribution during cardiogenic shock. J. Clin. Invest. 1983; 72: 935-947.

31. Macklem P.T.: Respiratory muscles: The vital pump. Chest 1980; 78: 753.

32. Roussos C., Macklem P.T.: The respiratory muscle: New Engl. J. Med. 1982. 307: 786.

33. Gordon E.E.: Anatomical and biochemical adaptation of muscle to different exercise. JAMA 1967; 201: 755.

Subject Index